# UNSUNG HEROES OF POLLINATION

*For my wife, Linda, and her Australian family. You gave me refuge and a sense of direction as we lived in Bundoora, Melbourne and Saint Louis.*

# UNSUNG HEROES OF POLLINATION

## More Than the Birds and the Bees

PETER BERNHARDT

PUBLISHING

A catalogue record for this book is available from the National Library of Australia

ISBN: 9781486319619 (pbk)
ISBN: 9781486319626 (epdf)
ISBN: 9781486319633 (epub)

How to cite:
Bernhardt P (2026) *Unsung Heroes of Pollination: More Than the Birds and the Bees*. CSIRO Publishing, Melbourne.

Published by:

CSIRO Publishing
36 Gardiner Road, Clayton VIC 3168
Private Bag 10, Clayton South VIC 3169
Australia

Telephone: [+613] 9545 8555
Email: csiropublishing@csiro.au
Website: www.publishing.csiro.au
**Sign up to our email alerts: publishing.csiro.au/earlyalert**

Front cover: Greenhood (*Pterostylis tunstalli*) and fungus gnat (genus *Mycetophila*) (photograph by Rudie Kuiter)

Edited by Lachlan Garland
Cover design by Cath Pirret Design
Typeset by Envisage Information Technology
Printed in Australia by Fraser & Jenkinson Pty Ltd



CSIRO acknowledges the Traditional Owners of the lands that we live and work on and pays its respect to Elders past and present. CSIRO recognises that Aboriginal and Torres Strait Islander peoples in Australia and other Indigenous peoples around the world have made and will continue to make extraordinary contributions to all aspects of life including culture, economy and science. The use of Western science in this publication should not be interpreted as diminishing the knowledge of plants, animals and environment from Indigenous ecological knowledge systems.

The paper this book is printed on is in accordance with the standards of the Forest Stewardship Council® and other controlled material. The FSC® promotes environmentally responsible, socially beneficial and economically viable management of the world's forests.

Feb26_01

# Contents

# About the author

Peter Bernhardt was born in Brooklyn, New York, in 1952, and lived most of his first 23 years in New York State. He went to the State Universities of New York at Oswego and Brockport for his Bachelor's and Master's degrees in biology. The Peace Corps sent him to El Salvador for two years where he taught botany courses and collected plants for the herbarium of the University of El Salvador. In 1977, he was given a research scholarship at the University of Melbourne and did his PhD in Victoria studying bird-pollination of box mistletoes (*Amyema*). He continued to work at the University in the Plant Cell Biology Research Centre on the pollination of wattles, commercial canola, guinea flowers and native orchids. Peter returned to America in 1985, married his girlfriend from Melbourne, Linda, and became a professor at Saint Louis University until he retired in 2019. He remains a research associate of the Missouri Botanical Garden (Saint Louis) and an adjunct professor at Curtin University, Perth. He became an Australian resident in 1992.

Peter is the author/co-author of over 100 scientific papers, and 23 books and book chapters. This includes popular books on plant life including *Wily Violets and Underground Orchids* (1989), *Natural Affairs* (1993), *The Rose's Kiss* (1999) and *Gods and Goddesses in The Garden* (2008). From 1990 to 1996 his column, 'The Botanical Detective,' appeared in *The Sydney Review.* He has also had articles published in *Natural History Magazine*, *Garden*, *Plant Talk* and *Good Gardening Australia and New Zealand.* He remains a foreign correspondent and commentator for *The Science Show* on Radio Australia.

In 2022, Peter received the Peter Raven Award from the American Society of Plant Taxonomists for outstanding contributions to public education in systematic botany. He remains a supporter of the Nutcote Museum in Neutral Bay, NSW, and the Kunming Institute of Botany in Yunnan, China.

# Acknowledgements

Much of the information provided in the following pages stretches over four and half decades as my career passed from student to educator, botanist and science writer. Several chapters are based primarily on field and laboratory studies in association with people with professional experience in their fields. Therefore, I must thank them starting with my long-term, research partner and photographer Dr Retha Edens-Meier. The chapter on wasps and *Paeonia brownii* would not have been possible without Dr Nan Vance, then of the USDA Forest Service. My earliest introduction to the winter orchids of Australia was through the generosity of Dr Malcolm Calder, my former PhD advisor. Later work in Australia on midge orchids and their eye flies, was arranged by Dr Peter Weston then of the Royal Botanic Garden Sydney and through the superior observations of research partner, Dr Wendy Grimm, whose family also provided essential ground support. Field and laboratory studies in Yunnan would have been impossible without Dr Zong-Xin Ren, Professor Wang Hong, and Professor De-Zhu Li of the Kunming Institute of Botany and their graduate students. For much of my career I was in debt to the guidance of Dr Leonard Thien, late of Tulane University.

All writers of nonfiction need to confer with people smarter than ourselves, to make certain we find the appropriate evidence and interpret it correctly. For this reason, I am especially grateful to Professor Robert Raguso (Cornell University) for his guidance on how floral scents are made and to Dr Roman Kaiser for sending me a copy of his magnificent book, *Scent of the Vanishing Flora*.

The manuscript of this book was written at the Missouri Botanical Garden. I depended on the efforts of the librarians in the Garden's Peter H Raven Library to retrieve books and research papers, and am grateful to Ellerman Book and Paper Conservator, Susie Cobbledick, who copied images in old volumes. Wildlife photographer Rudie Kuiter has spent far more hours observing the winter and spider orchids of Victoria, Australia, than I have and I am dependent on him for his descriptions and images. More recently, my reintroduction to the wiles of moths dependent on boronias is based on the ongoing studies of Drs Douglas Hilton (the Walter and Eliza Institute of Medical Research), Liz Milla (NCMI, Black Mountain) and Professor Axel Kallies (University of Melbourne).

Let's not forget the people who wake up your brain and make you start thinking about what you think you know and why it's time to rethink it. Dr Peter Raven of the Missouri Botanical Garden has been filling my need for a botanical guru since I was a Peace Corps volunteer in El Salvador in the mid-1970s. More recently, I've been stimulated by the emails, telephone calls and infrequent visits of Joey Santore, known to many through his YouTube channel, *Crime Pays But Botany Doesn't*. Associate Professor Nathan Muchhala, of the University of Missouri in St. Louis, gave me access to his photos and videos of bats and flowers based on years of fieldwork in the Neotropics. When it is time to explore the history of natural history in Australia there is no better place to go than the decades of publications produced by what is now the Field

Naturalists Club of Victoria (FNCV) and I am grateful to the editors for allowing me to reprint some of their classic illustrations.

A few field biologists are lucky. They live their professional lives in parts of the world in which the organisms and interrelationships they want to observe are available within a short drive or a brisk walk, and the cost of travel is negligible. When these conveniences are absent, you need long-term plans, generous funding and understanding hosts. For this reason, I must thank Dr Andrew Huber for access to GROWISER, a must-see wildflower reserve in Pumpkin Ridge, Oregon and to the faculty and staff of the Kunming Institute of Botany in Yunnan. Finally, it's long past time for me to thank the University of Melbourne for their 1977 PhD scholarship and two subsequent fellowships.

Since 1990, my field and laboratory work, relevant to this book, was supported financially by the National Geographic Society (America), the USDA Forest Service, the Australian Orchid Foundation and the Chinese Academy of Sciences (PIFI).

# Introduction: Another abominable mystery

*See the Children of earth, and the tenants of Air,*
*To an evening's amusement together repair.*

*And there came the Beetle, so blind and so black,*
*Who carried the Emmet, his friend on his Back.*
*And there was the Gnat, and the Dragon-Fly too,*
*With all their relations, Green, Orange, and Blue.*

*And there came the Moth, with her plumage of down,*
*And the Hornet, with Jacket of Yellow and Brown.*
*Who with him the Wasp, his companion, did bring,*
*But they promis'd, that Ev'ning, to lay by their sting ...*
– The Butterfly's Ball, and the Grasshopper's Feast, *William Roscoe*

The American Museum of Natural History in Manhattan mounted a large screen over one of its exits. Every few seconds it flashed a close-up of a different insect. The photo of the insect's body was followed by a second shot of its head, magnifying the eyes, mandibles and antennae. I found the screen by locating the voices of children. Boys and girls from one of the public schools were watching the big slide show. Every time the portrait of an insect's face appeared on screen, they shouted 'eew' in unison. Even butterflies weren't spared. I listened to the children for half a minute and then looked at their young teacher. She looked back at me with a bemused but guilty expression, as if to say, 'Don't blame me, and what else did you expect?' This was in 1977. If a public museum set up the same screen and showed the same slide show today, do you think the responses of most kids' would be any different?

I'd grown up as a would-be naturalist in an overbuilt, pesticide-sprayed, post-war suburb on Long Island, New York. The AMNH became one of my sources of knowledge and inspiration from boyhood through my university years. Our relationship became mutualistic. I donated specimens to their Entomology Department collected in El Salvador during two previous years in the Peace Corps. That day, this pilgrim made one last visit to his shrine before his departure that August. Thanks to a generous scholarship, I would go to the University of Melbourne in Australia, to conduct more field and laboratory work towards a doctorate.

My PhD years in Victoria were spent among songbirds that pollinated native mistletoes (*Amyema*). After graduation, I continued to work in the university's Plant Biology Research Center on the insects that pollinate many species of Australian wattles (*Acacia*). I was a member of the same team that found that the sunshine wattle (*A. terminalis*), was pollinated by a combination of birds *and* bees. After I left Melbourne, I spent an additional 35 years observing and glorifying the activities of dozens of bee species we now find so precious and life affirming.

What career could be more wholesome? Ah, but during my last years at Melbourne University I corresponded with Leonard Thien (1938–2021), a professor at Tulane University in New Orleans. He made me a member of his team working on a tropical project that some of you might find uncanny. During spare moments in the laboratory I was gold plating the tiny *Sabatinca* moths Len first collected in flowers he found in New Caledonia. The technician in charge of the scanning electron microscope, owned by the School of Botany, allowed me to record and identify pollen grains attached to the moths' wing scales. These insects ate pollen, instead of drinking nectar, and so preferred the fruity scent of certain flowers that male and female moths used as a public stage on which they courted and copulated. In the Australian slang of the early 1980s, the flowers of *Zygogynum* trees served as *Sabatinca* sin-bins. That is how Len, my friend, co-worker and mentor, introduced me to a side of pollination unknown to my hero Charles Darwin (1809–1882).

The study of pollination was young during the days when Darwin observed and experimented on British wildflowers. Today, pollination biologists like myself regard him as our touchstone. Darwin explained how floral adaptations turned insects into pollen taxis and how such mechanisms were derived from ancestral forms. Unfortunately, Darwin couldn't explain the origin of flowers as there were no specimens of fossilised blossoms available in his day to compare with the living. In letters, he referred to the origin of our flowering plants as 'the abominable mystery'. The rural countryside of what is now a part of South London was never the best place to look for model species that would lead him to sustainable theories. Len was willing to look elsewhere.

As a pollination biologist working within the second half of the 20th century, Professor Thien preferred fieldwork in quaking bogs and steamy forests to the safety of greenhouses and hay meadows. His research took him to Louisiana bayous, Australasia, Mexico and even to Madagascar. He observed and experimented on flowers belonging to obscure and woody species known mostly, if at all, from pressed specimens, glued to sheets, in museums owned by botanical gardens. Len wasn't much interested in the sort of floral displays that catered specifically to pretty butterflies, bees or hummingbirds. Most of the insects he caught or photographed were flies, beetles and small fuzzy moths. They were never the fauna that beguiled tasteful society in Darwin's day or our own. If they enter our houses today, we exterminate them.

When the Botanical Society of America asked one of Professor Thien's former students and myself to write an 'In Memoriam' for Len it was the least I could do. Tulane University, his widow Lorraine, two of his former graduate students and the Internet provided me with all the information we needed. Len was adept at teasing out the secrets in flowers by observing the behaviour of their pollinators. Then he used the information to interpret trends in evolution. If you are willing to come with me, you will meet him, and one of his orchids, in Chapter 3. Without his influence, and other great intellects like him, I'm sure I would have regarded my previous 33 years of employment as the deserved fate of another third-rate professor in a fifth-rate American university.

Re-reading Len's list of publications triggered bouts of free association in my head for weeks. I started remembering things from childhood including *The Ugly Bug Ball*, a song from a 1963

Disney movie when 9-year-old Peter had a crush on Hayley Mills. In high school, I'd read Bill Ballantine's, *Nobody Loves a Cockroach* (1968) which insisted we despise all urban wildlife and regard pesticide chemists as saviours. Such fragments reminded me that, while we've convinced ourselves we are far more serious about conserving the diversity of life today, the truth is that we are still biased regarding which species are worthiest of our protection. We claim to be stewards of the planet, but have you ever read EO Wilson's 1987 essay 'The little things that run the world'?

I've devoted years of my professional life to the study of those little things, particularly those interactions between plants and animals that would still encourage children to yell 'eew' and make many adults cringe. Working with other scientists who studied members of the iris family in South Africa I examined scores of hairy beetles acting as self-employed couriers of plant sperm (Chapter 9). I watched wasps guzzle nectar on three continents (Chapters 6 and 7). During a field trip to eastern Australia, I observed native orchids with pollinators even tinier than Len's moths. However, any local stockman or grazier viewing these flies would have found them pernicious (Chapter 5).

I want to use the study of pollination to address two overlapping issues. First, Australians probably agree it's past time for people all over the world to stop thinking of our corner of the Pacific as the centre of abnormal Nature. We're tired of more than two centuries of well-meaning writers expressing their amazement at the Austral freak show compared to the 'normal' flora or fauna living in their own neighbourhoods, thank you very much. Viewing any unfamiliar species as a jumble of endearingly odd traits is limiting unless we follow through and show how these traits allow organisms to interact with different organisms in order to survive. There will be majesty, complexity and yes, similarity in these interactions from Australia to North America, from South Africa to Fiji and on to China, provided we view them within a more holistic context. Using pollination systems that employ less familiar and often less-likeable animals will help take us there.

Second, what is a book of natural history without its natural historians? As in all branches of biology, the fields of evolution and ecology progress by uneven starts and stops. They often occur as sporadic torch races, in which previous generations pass their observations, experimental results and specimens down to us within old publications. To understand what we know about the science of pollination we need to meet a few of its pioneers and match them to those who are still working in laboratories and field sites to expand our perception of flowers and their partners.

To make my case, you and I must meet on common ground. The first chapter introduces you to the components of flower structure and plant chemistry most of us missed in school. In particular, what we know today is based in part on the recent discovery and description of fossilised flowers and flower-loving animals. Much of the most recent evidence occurs inside gemstones of amber. I think Darwin would have approved.

The second chapter introduces you to how major shifts in pollination systems may occur following minor changes to these floral traits coupled with changes in animal anatomy and foraging behaviour. Yes, let's begin by considering a lineage from its ancestral forms to its current descendants.

# 1

# Why and how do flowers vary?

*Fair pledges of a fruitful tree,*
*Why do ye fall so fast?*
*Your date is not so past,*
*But you may stay yet here awhile*
*To blush and gently smile,*
*And go at last.*

*What, were ye born to be*
*An hour or half's delight,*
*And so to bid good-night?*
*'Twas pity Nature brought ye forth*
*Merely to show your worth,*
*And lose you quite.*

*But you are lovely leaves, where we*
*May read how soon things have*
*Their end, though ne'er so brave:*
*And after they have shown their pride*
*Like you, awhile, they glide*
*Into the grave.*

– To blossoms, *Robert Herrick*

IMAGINE YOURSELF BUYING A BOUQUET FOR SOMEONE YOU LOVE. I LIKE shopping for carnations (*Dianthus*) in particular. I can choose from such a broad range of colours, pigment patterns (solid, streaky, blotched) and scents. There's additional variation to enjoy in the scalloping of the teeth or fringes of their petal margins. Now, imagine you are an animal and you eat pollen or the nourishment you want is sugar dissolved in flower nectar. Do you think you would select the same carnations you bought as a human for Mother's Day? As an insect, do you think you would be able to see the same colours, smell the same odours and feel the same petal textures you sense as a human? Do you think you would spend the same amount of time foraging on carnations compared to the time you passed shopping and purchasing them? Maybe those peonies in the next bucket will give you a better feed or what if you are comparing flowers to see which offers the best options for raising your children?

Plants never made flowers for us. The fossil evidence suggests that many generations of gene pools reacting to the pressures of natural selection have been sculpting, measuring, perfuming, painting and provisioning flowers for more than 100 million years. By comparison, the oldest man-like ancestors first appear in deposits less than 7 million years old. We call our species *Homo sapiens,* but that body form is only 300,000 years in the making. Flowering plants, on the other hand, not only survived the mass extinction events of 66 million years ago, they benefitted from them. They expanded their range, dominance and diversity following the initial cataclysms. Today, they literally grow over the graves of the biggest dinosaurs and the trunks of extinct, cone-bearing trees. Yes, *Homo sapiens* appreciates flowers and so do several unlikely creatures our civilisations regard with dread, and often for good reasons. In this chapter, we will see how flowers attract pollinators and that includes some animals some of us still prefer to avoid. And why do certain animals visit the flowers of some species but ignore others?

We obviously have some idea what attracts us when we buy a bouquet from a shop or when we linger in a garden. The problem begins when some of us get a little too close to flowers. The architecture and arrangement of their organs can be bewildering even when we look at them closely with hand lenses. The standard terminology for floral parts and shapes we use today has undergone expansion and refinement since the 18th century. It only looks daunting because most of us have either never been taught technical flower-speak or we've forgotten our school lessons because we never used them outside class, like trigonometry. I come from a generation of schoolchildren expected to spend a lot of time recognising words but their importance was relative. A third grade teacher taught us homonyms using pistol and pistil as prime examples. A pistil was a part of the flower according to the children's dictionary she left on our desks. I can't say I learned much more about the other organs in a flower until I majored in biology at university. Today, hundreds of thousands of university students still major in biology without ever taking a botany class, let alone the sort of zoology lectures that introduce you to animals that pollinate flowers.

How are we to recognise a flower anyway? Do they all have rings or spirals of

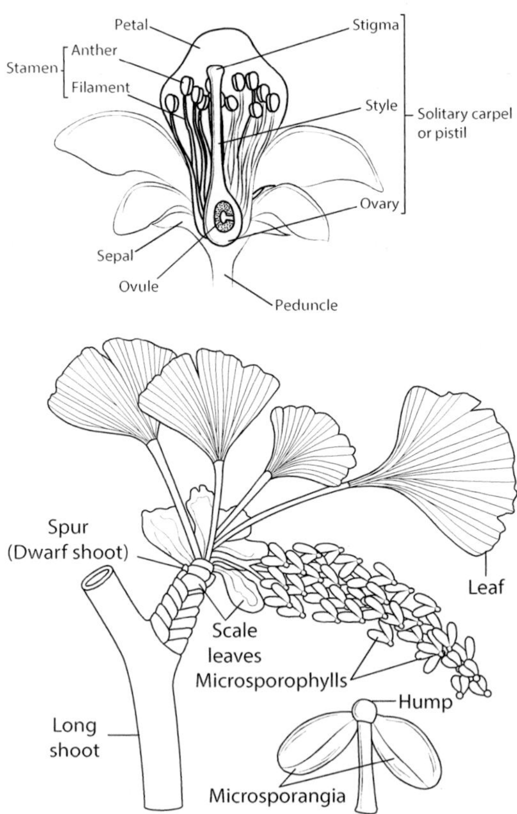


(Above) This generic half-flower is complete (four kinds of organs in four whorls), perfect (bisexual) and the pistil contains a single ovule inside the ovary of a single carpel. (Below) In contrast gymnosperms usually keep male and female organs on separate branches and *Gingko biloba* takes the separation of sexes to an extreme. The nasty smelly seeds come from female trees. Here we see the pollen-making 'cone' of a male gingko.

recognisable petals? Nope! It's a matter of packaging. Place the youngest reproductive stalks of cone-bearing trees and shrubs under a microscope during their earliest phases of development and you will find their exposed and unfertilised seeds (ovules) forming on the top of a structure resembling a flattened scale. Sometimes two ovules form on the same little scale exposed to the air while they wait for incoming pollen grains. Each ovule has a pore and oozes a sticky droplet that catches incoming pollen grains blown in on a breeze from a male cone on another twig on another tree. Real flowers don't do that and you wouldn't confuse flower ovules with those ovules protected by the hardened shingles of a pine cone or cycad, or dangling naked from a stalk on a branch of a female ginkgo tree.

True flowering plants also let their little ovules develop on a little scale. That scale keeps growing and folding until it is so large and hollow that it 'swallows' the ovules shutting them off from the surrounding air. The still-virgin ovules remain attached to the inner walls of the hollow box or capped bottle. When the flower opens, pollen grains have only one place to land and that's the glandular apex that makes up the bottle's 'cork'. This pollen-catching tip is called a stigma. While the stigma waits for incoming grains and the ovules hidden inside remain unfertilised we refer to the whole female organ as a carpel (immature fruit). If two or more carpels develop in the bud by fusing together so completely that they share the same cork (stigma), it's better to refer to the now compound organ as a pistil.

What was the original advantage of enclosing developing seeds inside a box or bottle? We will probably never know (and who says there was only one advantage?). The descendants of the first flowers do well with this innovation. We know that the interior of the ovary at the bottom of the modern pistil is a moist chamber that protects its developing ovules from arid atmospheres. The wall of the ovary also protects ovules from some destructive insects until they develop their own hardened seed coats. Once ovules receive pollen sperm the ovary ripens into a fruit wall. Ripe fruits offer new options for dispersing offspring away from their parent plant so they don't compete with their seed mother. Some fruits turn their ovary walls into glider wings. Others fill with air pockets allowing seeds to raft down rivers. The fruits of some violet species (*Viola*) open and then their drying valves squeeze the ripe seeds until the pressure shoots them into the air. Let's not forget all the fruits that turn one or more layers of ovary tissue into something eatable and rich in sugars, fats, vitamins and/or triglycerides. Fleshy fruits are carried off and consumed by animals that eventually release seeds intact when they poop or regurgitate. Were the first fruits ever dinosaur food? The great tropical botanist EJH Corner (1906–1996) teased his readers when he presented the image of the earliest toothed birds munching primordial berries.

Perhaps the greatest advantage to the closed carpel was the one that remained invisible to human eyes until Darwin and his cohorts began hand-pollination experiments back in the 19th century. The stigma surface and interior layers of carpel tissues form a selective passageway. It's up to the pistil to recognise then choose or reject the sperm-filled tube a pollen grain grows when it lands on a stigma. As the pollen tube attempts to pierce the stigma surface and grow down towards the ovules inside the ovary the pistil must 'decide' whether to keep or reject the penetrating tube. It's always the female organ in the flower that ultimately determines who sires its seeds. Based on his experiments and correspondence, Darwin concluded that 'Nature abhors perpetual self-fertilization'. Later generations of plant breeders understood that cross-pollination promotes genetic diversity among members of the same population. The same recognition system in a pistil also allows a parent plant to reject the pollen of alien species and avoid producing a generation of sterile or semi-sterile hybrids.

When did plants first make carpels or pistils? As usual, palaeontologists argue over who has the oldest fossil. All they will agree upon is that the earliest flowering plants grew in a part of the world we now call Eurasia. The oldest grew in freshwater and lacked true wood. If you side with the Chinese, you marvel over *Nanjinganthus* (flower of Nanjing). first described in 2018. It looks like it had a closed ovary to store its seeds and it may even have had petal-like organs that formed a circle on top of the ovary. Its stigma had many lobed branches to trap incoming pollen and looked like a miniature brush. The fossil dates to 174 million years old and that puts it in the Jurassic with early stegosaurus and brachiosaurus. It is nearly 25 million years older than the fossil of the bird-like dinosaur we call archaeopteryx.

Other palaeobotanists think this is a misinterpretation and do not accept that *Nanjinganthus* had an ovary that enclosed seeds. For them, the first flower fossils date from 125 to 130 million years ago, placing them in the early to mid-Cretaceous with iguanodon and the duck-billed ouranosaurus. They point to the reconstruction of the carpels of water plants like *Montsechia* from Spain or *Archaefructus* (ancient fruit), which is also Chinese. The *Montsechia* fossil is estimated to be at least 130 million years old. It didn't have petals and probably bloomed underwater like its modern relatives, the hornworts (*Ceratophyllum*). *Archaefructus* appears in younger rocks ~120 million years old. This water plant pushed its 'flowers' above the surface into the air and could have exposed its stigmas to visiting insects. The same exposed stalks also bore separate clusters of male pollen-making organs. Once again, *Archaefructus* had no petals and lacked the protective, bud-making organs we call sepals. Some people compare the reproductive stalks made by extinct *Archaefructus* to the modern flowering stems of black pepper vine (*Piper nigrum*) and its *Peperomia* cousins in the family Piperaceae. These living plants appear to be pollinated by a combination of the wind and small pollen-eating insects with short tongues like hover flies and solitary bees.

## ANCIENT TRENDS IN THE EVOLUTION OF FLOWER ORGANS

As we look at the interpretation of flower fossils from the early Cretaceous to the asteroid crash of 66 million years ago, we can pick out three important trends in flower evolution.

First, flowers start varying in size during the Cretaceous. While most remained small and were no larger than pinheads, we see that 100-million-year-old fossils of *Archaeanthus* (ancient flower) trees offered bigger blooms. Fossils of their wood cells, leaves and floral architecture show that they were related to modern magnolias (*Magnolia*) and tulip trees (*Liriodendron*) still surviving in temperate Asia and North America. Flower size appears to be a plastic trait when plants from the Cretaceous are compared to their descendantstoday. The reconstructed flower of *Microvictoria* (miniature *Victoria*) was smaller than your pinky nail. It is recognised as one of the ancestors of some members of the modern water lily family (Nymphaeaceae), now loved for their fat, full flowers. *Microvictoria* is extinct but its surviving relatives include three species of *Victoria*, the biggest plants in the family. *Victoria* flowers are the size of cabbage heads. Each massive bud always blooms and dies within less than 24 h. Increasing lifespan and physical size do not represent a united trend in the evolution of flowers.

Second, most of the other flowers of the Cretaceous stopped arranging their sexual organs in a continuous spiral along a thin stalk. A few, like the magnolias and custard apples (Annonaceae), continue this tradition to the present day, but add large and flat petal-like organs (tepals) to the base of the spiral. Ultimately, the magic wand of spirally arranged tepals, followed by spirally arranged male organs (stamens), terminating in a final helix of female organs (carpels) was replaced by the modern flat-headed flower. In the majority of flowers, the same organs are confined to their own discrete circles, known as whorls. The number of organs in each whorl is often reduced to only three or four, or they occur as increasing multiples of three, four or five.

Third, by 100 million years ago, the fossil evidence suggests that most flowers were limited to four kinds of organs, with each organ restricted to its own conservative, stereotypical whorl.

The bud case protecting the immature male and female organs is now made of sterile and often green, leaf-like sepals forming a whorl called the calyx. When the sepals open, they make all the organs in the remaining whorls look like they are arranged inside an ancient Greek, wine cup with handles, the kylix. Once the sepals open the next whorl to unfurl is composed of petals. As these are usually the largest and most brightly coloured organs in the flower it's no wonder we call this whorl the corolla (the little crown). Above the corolla is one or more whorls of male organs we call stamens. As stamens make pollen and pollen grains are microscopic boxes containing sperm, this whorl is known as the androecium (the men's apartments). Finally, the carpels or solitary pistil is the last and central whorl. As the ovaries contain ovules and each ovule must contain an unfertilised egg we call the whorl a gynoecium, after apartments meant only for women.

## DARWIN'S GRANDPA CONSIDERS FLORAL VARIATION

The development of four different organs in a flower and their arrangements in whorls has varied intensely over the last 100 million years. Most variation is natural with genes turned on and off in the course of evolution. Variation is also artificial as plant breeders once preserved rare mutations selecting for new forms. Today they induce them with modern genetic techniques. Flowers produced instead by natural selection vary from each other in at least three ways and each prospective pollinator must adjust to them. For example, genetic variation allows some flowers to eliminate entire whorls. Grasses (Poaceae) are pollinated almost entirely by currents of air. That's why we usually do not let them flower in lawns. The proteins worn on the pollen grain walls help stigmas recognise them, but they become allergens when we inhale them. As the wind does not prefer a particular colour or scent, grasses have eliminated the whorl of petals. In the 18th century a flower lacking one or more whorls of organs was classified as 'incomplete'.

Not all flowers are bisexual either. When a bud made by a begonia (*Begonia*), marijuana (*Cannabis*), papaya (*Carica*), pumpkin (*Cucurbita*) and many more opens, it contains only functional stamens or pistils. Sometimes male and female flowers share stems on the same plant, as in begonias and pumpkins. Sometimes there are separate male and female plants in the same population or crop as in marijuana. By the standards of 18th century botanists, a unisexual flower was more than incomplete. It was also imperfect because these scientists didn't understand the importance of cross-pollination. If there were two sexes in each flower, earlier botanists believed that Sir Stamen married Lady Pistil. Of course, with so many stamens and so few pistils in the same flower a polyandrous wedding day was expected according to the plant taxonomist Carl Linnaeus (1707–1778) and his loyal poet of botany, Erasmus Darwin (1731–1802), who was Charles Darwin's grandfather. Erasmus took Linnaeus' metaphor that the flower was a bridal bed just a little too literally. He would take his hand lens and look inside the pink and white flower of an autumn crocus (*Colchicum*) comparing it to a nymph out of Greek mythology. In the centre of the flower, the pistil's neck subdivides into three long, female lobes (styles) and all three face an outer circle composed of six male stamens. Without charging the pistil with bigamy, Erasmus wrote ...

*'Three*
*Blushing Maids the intrepid Nymph attend*
*And*
*Six*
*Gay Youths enamour'd train! Defend'*

And remember, gay youth had a different connotation back in the 18th century. We don't bother with the euphemisms of Erasmus Darwin anymore. The true vocabulary of flowers, while unfamiliar, has its own logic and with scientific evidence comes a unique beauty.

## FLORAL FUSION AND FLOWER SYMMETRY

There's a second way that flowers vary, and thanks to the fossil evidence it's clear this development began in the Cretaceous and continues to the present day. When each flower organ begins to develop it starts life as a microscopic bump of rapidly dividing cells. As organ development is so crowded inside the same bud, what do you think happens when bumps in the same whorl fuse together? We've already noted that when little carpel bumps fuse together you end up with a giant pistil. When sepals or petals fuse, they unite side by side as if they were children holding hands in a circle of *Ring around the rosy*. This produces the bells, trumpets and tubes of flowers we love today. When like fuses with like in the same flower we call the process connation. The corolla of a primrose (*Primula*) forms a funnel of five connate petals.

Let's take it one step further – bumps fusing with bumps belonging to different whorls. This explains why stamens in a mint bush (*Prostanthera*), snapdragon (*Antirrhinum*) or grevillea (*Grevillea*) are attached to the funnel, tube or pouch made by the connate petals of the corolla. When organs in one whorl fuse to those in a different whorl, we call the process adnation. Take another look at the same primrose. Those stamens are now adnate to the throat of the corolla. This mode of fusion can be so extensive it can alter the entire architecture of a flower. When sepals are adnate to petals, and petals are adnate to stamens, and stamens are adnate to carpels in the same flower, the pistil's ovary becomes buried under so many layers of lower tissue that they all become part of the fruit wall. Turn an apple or pear upside down and look at the scabby dimple in the centre. Can you see the remains of the sepal flaps and little bristles that were once stamens?

In general, the way in which closely related plants determine the number of organs in a whorl and how they fuse together looks so similar and can be so predictable they offer reliable guides when assigning extinct species to appropriate twigs on a modern family tree. Just as we assign the skull and canine fangs of an extinct sabre tooth tiger to the cat family (Felidae), we can reconstruct some fossil flowers and their petrified pollen grains and assign them to a modern plant family. That is how palaeobotanists attempt to reconstruct the vegetation of Cretaceous forests and lakes before the asteroid hit 66 million years ago. They identify ancestors of magnolias (Magnoliaceae), avocados (Lauraceae), mountain pepper (Winteraceae), water lilies (Nymphaeaceae), saxifrages (Saxifragaceae), plane trees (Platanaceae), and perhaps, the first palms (Arecaceae). By the end of the Cretaceous, trees with reduced, wind-pollinated flowers included ancestors of willows (Salicaceae) and oaks (Fagaceae). Of equal, if not greater importance, there are flower fragments from the Cretaceous found in the fossilised tree resins we treasure as the gemstone, amber. Many of these extinct blooms can't be assigned specifically to extant families. Did they become extinct long before humans existed, or are we looking at something more formative? Could we be looking at the ancestors of modern Orders that have not yet diversified into more than one discrete family? As we will see, amber also gives us pollinators to ponder.

Let's look at the third source of physical variation in flowers. We return one last time to those microscopic bumps of rapidly dividing cells. Different bumps grow and expand at different

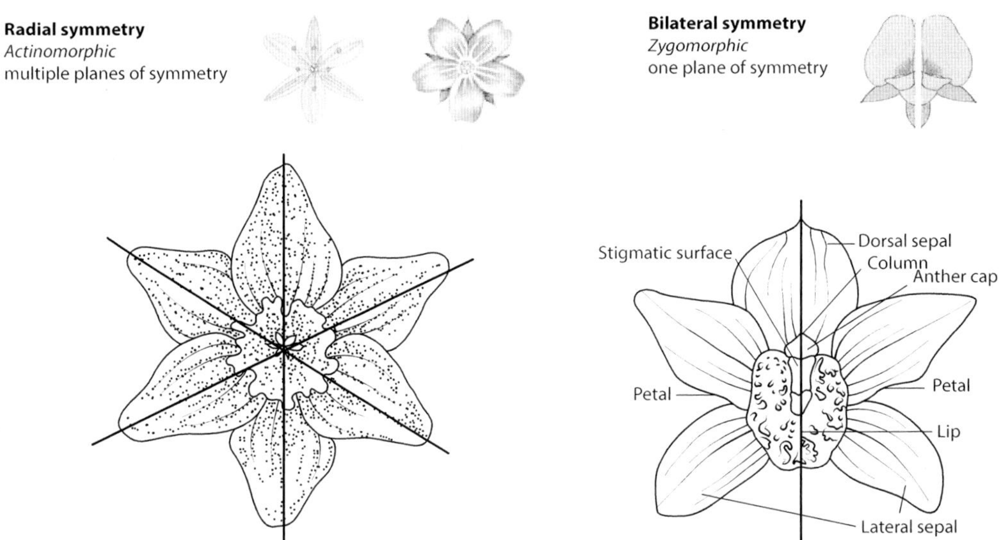


Radial versus bilateral symmetry. Note how a radially symmetrical flower can be divided into two halves at any angle. A bilateral flower can only be divided once longitudinally to produce two mirror halves. Illustration by Envisage Information Technology, with three small flowers adapted from Enid Mayfield's *Illustrated Plant Glossary*, 2021.

Extreme bilateral symmetry of a hammer orchid (*Drakaea*) of Western Australia showing exaggeration of the size, shape and sculpturing of its lip petal versus its two skinny lateral petals and three skinny sepals. Photo by Mark Brundrett.

rates even within the same ring of organs. Many plants are programmed to produce petals of different sizes and shapes in the same bud. This ultimately changes the symmetry of the flower.

If the flower looks like a bell or a bowl, it means that all the bumps in the corolla and/or calyx developed at the same rate and are all the same size and shape. We can compare this uniform corolla to the radial symmetry of a sea star in which all the arms are the same length and width. Now, compare this to toadflax (*Linaria*), violets, impatiens, most orchids and the flowers produced by an Australian mint bush (*Prostanthera*) or their garden relations like *Salvia.* Even when their petals are fused together (connate) the free lobes of some petals are larger and differently shaped on one side of the flower compared to the other. Like human beings, such flowers show bilateral symmetry. They have a left and right side, and you can divide them in only one direction to produce two equal, mirror halves.

## DOES EVERY GOOD POLLINATOR DESERVE FAVOUR?

All flowering plants must make fertile seeds to produce new generations. The fact that these flowers have varied, and continue to vary in the number and kinds of organs, physical size, degrees of fusion and symmetrical shapes indicate that different flowers appeal to different pollinators at different times. Before we ask the important question of which animals were pressed into the role of pollen chauffeurs in the Cretaceous, we must ask another logical question. Why would any insect enter a flower in the first place, especially if the flower forces an animal to hang upside down on a stem, offer only one landing pad or is so narrow that the creature can't push its entire body into it but must hover in the air or perch on a petal lobe?

The answer is that, in most cases, there must be a reward, or as the next chapter shows, at least the false promise of a reward. In other cases, the reward skips the adult generation and a part of the flower becomes a nursery to bring up the kids (Chapter 9). In most cases, though, flowers are pubs and you receive something nice to drink and/or eat. If you are a serious patron of a particular cuisine, you stick with a group of flowers offering dependable rewards in the same season. This leads to cross-pollination when flowers on one plant run out of menu items and its customers must find another establishment offering the same drinks and snacks. Remember, in most cases flower foragers aren't personally committed to transferring pollen from the stamens in one flower to the receptive stigma in another. Pollination is merely the happy and consistent consequence of a pollen-dusted visitor making passive contact with stigmas, although some shrubs and wildflowers as discussed in Chapter 9 are very important exceptions to this rule.

Pollen was probably the earliest edible reward if we go back far enough in the Cretaceous. While it's odd to think of animals actively eating sperm-filled particles we must understand that at least a quarter of the contents inside a pollen grain fed to pollinators is nutritious, containing amino acids, fats, vitamins, starch and some minerals. But there's one problem. Some modern animals are incapable of removing the nutrients inside the grain. The wall of a pollen grain has a double layer and that explains why pollen fossilises so well. The outside of most grains forms a sculptured wall spun out of a natural plastic based on chains of Vitamin A molecules. We all know how long it takes for commercial plastics to degrade. To make digestion even more difficult, the same pollen grain has a second wall under the plastic wall and it is made of cellulose. Most pollinators can't digest either wall or even break through them. Those that can, like the beetles we will meet in Chapter 8, have special mouthparts to manipulate, crush and swallow them. Bees are major pollen eaters because they have the most specialised digestive tracts adapted to swallowing whole grains that are hydrated until they rupture and then their nutrients can be extracted. Flower-pollinating bats (Chapter 10) have similar equipment, but the remainder of animals in this book can't digest pollen. What do most flowers offer as rewards instead?

## NECTAR FOR THE MASSES

Returning to the Cretaceous, let's look inside the reconstructed cups of flowers that blossomed at least 100 million years ago. We come to the conclusion that some were secreting the first nectar, a watery fluid enriched with sugars made first by the plant's green leaves during the act of photosynthesis. Sugars can be moved from one part of the plant to another by linking strands of conducting tissue called phloem. The sugar will be diluted further with water kept in additional storage cells. There are two ways to push the nectar out of the flower. Since the 1600s hand lenses showed that many flowers had little, weeping glands called nectaries. They are fixed to the skin of different organs in the flowers of different species. Early botanists mapped their positions with confidence, as the location of a nectar gland in a flower was an inherited trait that did not change on whim. The release of nectar from each gland differed according to species. Some squeezed it out through damp, surface hairs. Some had depressions or gutters

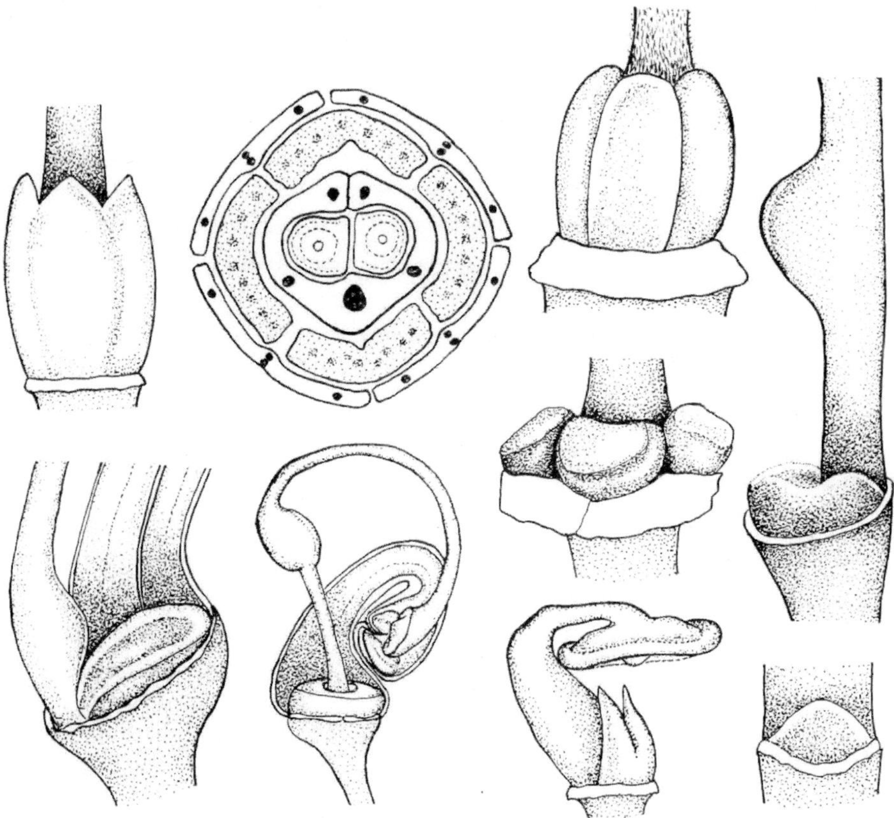

When flowers in the banksia family (Proteaceae) have nectar glands they are always at the base of the ovary but occur in different shapes, sizes and numbers according to genus. Top row from left shows the tall pointed crown of *Lambertia formosa* followed by a transverse section of the nectar gland surrounding the double ovary of a *Dryandra nivea* while a tall, lobed gland of *Austromuellera* is to its right. This is followed to the far right by the one-sided cup-shaped nectary under the elongated ovary stalk of *Finschia chloraxantha*. To its left, in the middle are the pebble-shaped nectar glands of *Lomatia tinctoria.* The four bottom drawings (from left to right) are the bathtub gland of *Grevillea* (*Hakea*) *crassifolia*, the smooth donut nectary of *Grevillea leucopteris*, the snake-tongued nectary of *Opisthiolepis* and the reduced scale gland of *Oreocallis*. Adapted from the 1976 paper by C. Venkata Rao in *New Phytologist.*

that overflowed at specific times like drainage causeways. In some cases, the glands deteriorated at maturity releasing the drink in a slush of deflating cells. Other glands are filled with holes and water pressure in the plant body, which makes them leak like sieves.

The second but less common mode of nectar secretion was found by the Austrian botanist Stefan Vogel (1925–2015). These structures can be found only with a microscope as they don't form definitive glands and they don't have discrete, stereotypical locations on specific floral organs. Instead, these domed collections of skin cells explode releasing sugar water or they simply form borders surrounding modified micropores known as stoma. Vogel called these cells nectarioles. Stoma are far more common on the under surfaces of green, photosynthetic leaves. The most important functions of leaf stoma are to exchange gases with the surrounding atmosphere and release excess water vapour.

What do real nectaries and nectarioles have in common? They are positioned towards the base of the flower. As the animal pushes its head in to drink it brushes against open stamens releasing pollen and/or the pistil tips (stigmas) waiting to receive pollen. The animal will carry off sticky or spiny grains once it finishes drinking. In its search for nectar in a second flower the pollinator brushes against or bonks into the stigma completing the process. There are all sorts of variations on this theme and the nutrients vary. Some flowers in the locust berry and Amazon vine family (Malpighiaceae) replace nectar glands with oil glands (elaiophores) on their sepals. Bees whip up the oils, mix them with pollen and feed their offspring a mayonnaise and pollen remoulade.

## REWARDS MUST BE ADVERTISED

The next and obvious question is how do animals know when the flower is offering any sort of reward? This is an answer we can't interpret from fossils as colour patterns and scents do not survive fossilisation. The easiest explanation is that floral scents first alert animals from a longer range, while the contrasting colours on the flower's skin indicate the best places to land and forage in the short range. It also explains why we, as primates with good colour vision but a limited sense of smell, like certain flowers yet ignore or are repelled by others. The same may be said for the thousands of animals that pollinate flowers.

Why do flowers vary in scent and colour? As fossils can't answer that question we must depend on experiments performed from the 20th century to the modern day that combine our knowledge of plant biochemistry with studies of animal behaviour. The short answer is that flowers vary in scent and colour because the senses of different animals vary in strength and definition. People who study animal behaviour refer to this as the umwelt. That's the part of the environment an animal can actually perceive using its own sensory organs and then its nervous system processes and interprets incoming stimuli as cues. As the senses of sight and smell differ so much across the animal kingdom, and even among closely related creatures, natural selection often favours plants that make flowers appealing to the most common and efficient pollinators in the environment. However, as the size and diversity of flower-visiting fauna also vary from continent to continent and over long periods of time, some plants have very few but very specialised pollinators with exaggerated senses.

Umwelts vary widely across an animal kingdom full of flower visitors and mere flower fanciers. For example, while red to orange petals and stamens often attract the attention of certain birds, long-tongued flies, butterflies and some beetles (Chapter 8), most bees remain unimpressed. Native plants with red to orange flowers abound in the tropics and warm temperate zones of the Southern Hemisphere becoming less frequent in the northern floras of Europe and Canada where such flower-pollinating birds and specialised insects are less common. Likewise, humans see many flowers as white, but many pollinators target blanched petals as they can see intricate patterns in the ultraviolet range, which we can't see. Similarly, we can't smell the blossom of an Australian bearded orchid (*Calochilus*), but certain male wasps find the fragrance irresistible, even when masked by blades of invasive grasses (Chapter 7).

Of the two major attractants, the bases of flower colour are better understood. Almost every hue or pattern we see in flower skins is based on only four major classes of chemicals. To our eyes, some chemicals make the same colour. For example, the betalains are a group of water-soluble acids and their molecular skeletons carry nitrogen atoms. We see them in the red of the carnation's petals I bought earlier and the skin of the dragon fruit (*Hylocereus*). In contrast, the red in a ripe apple's peel or the flower of a Christmas bell (*Blandfordia*) is determined by anthocyanins. They are also water-soluble and form short phenolic chains lacking nitrogen. Their intensity changes from pink to bloody scarlet according to the degree of natural acidity in the plant's skin cells.

Regardless of pigment group, plants can produce special effects by layering one coloured cell on top of a second provided it contains a different pigment. However, there are no black flowers. The jet-coloured dots or streaks on pansy petals (*Viola tricolor*) and their relatives are based on arrangements of a layer of blue cells positioned over a layer of deep purple made out of two kinds of anthocyanins. The vividness or texture we see in coloured flowers is often related to how skin cells are shaped, varnished and refract sunlight. If the cells are short, smooth and have a thick, transparent waxy cuticle it will give them a shiny-polished tone. This effect makes the yellow corollas of wild buttercups (*Ranunculus*) look like they were glazed in a kiln. If the cells are pointed and the waxy coat is applied in unequal layers, the petals seem velvety or furry.

Scent glands share one thing in common with nectaries. They are also collections of glandular cells fixed to flower skin, but they differ in the following ways.

- They don't appear to connect directly to the sugar conducting veins inside flower organs.
- They don't secrete liquids.
- Scent molecules are so volatile they burst into smaller, airborne particles at room temperature.
- Scent glands can be harder to locate on a flower than nectar glands because they are not restricted to the base of organs. The flowers of some wild roses scatter scent glands on their sepals, petals and stamens, and possibly on their pistils. They are so small you can only find them by sectioning and staining the skin of each organ. In contrast, some members of the orchid, philodendron (Araceae) and frangipani (Apocynaceae) families mass them together on different organs until they look so big, swollen and dense they resemble clubs or warts. When scent glands are so massive that they are visible to the naked eye, we often call them osmophores.

Different organs on the same flower may bear scent glands that make different combinations of odours. Pollen often smells different from whole petals. Most pollen grains are released with droplets of oil on their walls. This 'cold fat' is the perfect matrix for lightweight scent molecules much as cold pressed olive oil holds the odours of garlic or chilli.

How did flowers acquire their perfumes? Remember, all flowers are merely compact branches and their organs are all modified leaves, as in the last stanza of Herrick's poem. Leaf skins are natural factories producing warning odours, even toxins, to drive off grazing pests. We like the smells of leaves of culinary herbs but many insects do not. In New Caledonia, flowers of the *Zygogynum* trees release scents unusually rich in ethyl acetate. At higher concentrations the chemical would kill insects. Modified and mixed with oxygen it has a pleasant candied fruit smell and *Zygogynum* species attract small moths that find the flowers a perfect place to mate, as mentioned in the Introduction.

During the last years of the 20th century and into the first two decades of the 21st tremendous advances were made to our understanding of how wildflowers mixed perfume notes. Fragrance chemistry is often far more diverse than colour in the same flower and very different plants can produce and exploit different concentrations of the same scent molecules. When our team captured the odour of the flowers of the Australian bitterleaf vine (*Trimenia moorei*), analysis showed only two dominant molecules, 8-Heptadecene and 2-Phenylethanol. While the bitterleaf vine is among the few plants to employ 8-Heptadecene as a floral attractant, the same can't be said for 2-Phenylethanol. That molecule is mixed in the natural fragrances of dozens of unrelated species including the flowers of some orchids, magnolias, water lilies, and twinflowers (*Linnaea*). This suggests that it's a scent note that evolved early on in the evolution of at least one of the first odourousflowers and the gene or genes remain active in many descendants. Why? A surprising number of insects find 2-Phenylethanol innately appetising. When certain butterflies, beetles, stink bugs and lacewings are exposed to it, their first response is to reflex their antennae and then stick out their 'tongues,' as if anticipating a lovely treat.

Think of a scent molecule as a note in a piece of music. Very different plants compose their scents using the exact same notes but in very different chords and combinations.So 2-Phenylethanol or ethyl acetate is played *forte* in some flowers and *pianissimo* in others. The Swiss Biochemist Roman Kaiser analysed the fragrances of hundreds of flowering species, many in danger of extinction. He trapped the natural perfume of *Franklinia alatamaha*, a famous member of the camellia family (Theaceae) named in honour of Benjamin Franklin. The franklinia vanished from American forests after 1803 (no one knows why) and its future depends exclusively on gardeners. Almost 45% of its scent is based on methyl benzoate. In its pure state, this chemical reminds some people of ripe fruits of commercial feijoas or guavasteens. That's a pleasant aroma, but what do we make of the 46 other molecules mixed up in the same perfume ranging in concentrations from less than 1% to over 5%? Are they mere byproducts of the scent-making process or do they also play a subtle role in alerting appropriate pollinators? It's important to note that people smelling a franklinia flower for the first time are more likely to compare its charms to orange blossom (*Citrus x sinensis*) or Japanese honeysuckle (*Lonicera japonica*), not fruit. To humans, at least, very different flowers often smell alike.

Just as more than one gene controls the range and patterns of colours in the same flower, they also influence how it smells. Flowers of closely related plant species often have different smells to the human nose. Dr Kaiser must like South African freesias in the gladiolus family (Iridaceae), as he analysed the scent of 13 species. One species released as few as 13 scent notes, while another secreted 50. *Freesia andersoniae* was rich in linalool (92.3%), while *F. occidentalis* was dominated by nerol (77.8%).

Kaiser's work also showed that scent composition differed within the same species and location may have a lot to do with it. He trapped the fragrance of *F. caryophyllacea* near the

Bot River in South Africa. The dominant scent peak (35%) went to (E)-Ocimene. At the Kirstenbosch National Botanical Garden, a pampered bloom of the same species pumped out 34% concentration of alpha-Terpineol instead. The suggestion here is that natural selection may alter flower scent compositions over several generations when pollinator diversity and visits change from site to site. As we will see in Chapter 9 on boronias, the perfume and flavour industry is well aware that plants belonging to the same species, but restricted to isolated populations, may smell differently across their natural range and this can be exploited when they are brought into cultivation. When plants in different populations or their man-made varieties are analysed their scent contents can be pooled and the same species may yield over 100 different molecules; however, some notes are more distinct than others (Chapter 9).

Some of the most recent discoveries suggest that the intensity of some flower colours and some scents may link to the same common molecules. Many unrelated, but night-blooming plants with white, pale green or light yellow petals have similar fragrances that delight us because their piercing scent notes overlap. Natural perfumers like tuberose (*Polianthes*), cacti with nocturnal flowers, jasmine, evening primroses (*Oenothera*), gardenias, some tobaccos (*Nicotiana*), and the lady of the night orchid (*Brassavola nodosa*), open fresh buds from dusk into the witching hours. They are most likely to blend three classes of chemicals, but not in equal volumes. The acyclic terpenes offer fresh citrusy notes. The aromatic esters remind us of culinary herbs or medicinal extracts. We would find this combination too intense and cloying if not for the added depth provided by nitrogen-rich molecules that add an unclean, animal-like odour as a counterbalance.

## WHAT WERE THE FIRST POLLINATORS?

All of this variation in floral shape, size, rewards and attractants indicates that the selective foraging of different animals was, and is, a driving force in flower evolution. Different animals select different flowers to nourish them and/or their offspring. What can this tell us about the identities of the earliest pollinators of flowering plants during the age of dinosaurs? Until recently, botanists were uncertain how to assess and interpret the fossil specimens of flower-visiting insects. It was easiest to push the beetle theory and it went something like this. Beetles are among the oldest groups of insects with their earliest fossils at least 400 million years old ... right? They are also among the oldest surviving insects, dependent on seed plants and their products to complete their lifecycles ... right? Several surviving but 'primitive' plants like magnolias, custard apples, water lilies and even some cone-bearing cycads are still pollinated by beetles, so these insects must be the first pollinators ... right?

Let's backtrack for a moment. The first two decades of the 21st century were good ones for the discovery, reconstruction and interpretation of flower and insect fossils from the Cretaceous period. Specifically, this meant paying more attention to the so-called Burmese amber of Myanmar estimated to have formed over the previous 100 million years. In contrast, the most popular ambers for jewellery since Greco-Roman times came from the Baltic region and are no older than 48 million years old. Likewise, the fashionable ambers of the Dominican

Republic available after the mid-20th century were deposited only 25–16 million years ago. These younger ambers are more likely to reflect far later innovations among animals and flowers, and belong to a more recent past in which large mammals, modern birds and flowering trees increased their domination of life on land.

An amber fossil of *Melittosphex burmensis*, a wasp-like bee of the Cretaceous. Photo by George Poinar.

Let's be excited but prudent when we consider what's inside these much earlier jewels from Myanmar. It's here Chinese paleontologists found the 98-million-year-old remains of *Pelretes vivificus*, an ancient short winged, flower beetle in the Kateretidae, a family still distributed throughout much of the world. Most importantly, the fossil includes grains of pollen around the beetle's jaws and between the hairs of its brushy abdomen. Each grain had three grooves on its round wall indicating it belonged to an extinct member of the largest group of flowering plants on earth today, the eudicotyledons. Other beetle remains are known from this time period. Indeed, we have beetle coprolites with identifiable pollen in their petrified poop. Does this mean they were the first pollinators?

I'm afraid not. Drs George Poinar of Oregon State University and Bryan Danforth of Cornell University have found and described several bee-like beasts in amber from Myanmar. The insects' bodies have some waspish traits and their mouthparts are rather short, but they have brushy branching hairs on their hind-legs and that could only mean that the females collected pollen. You can argue whether these insects represent an extinct family of bees or they are something more intermediate, like a *basp* or a *wee*. In either case, they are placed in two new genera, *Melittosphex* and *Discoscapa*. Further examination of a female specimen of *Discoscapa* showed that its leg brushes were packed with pollen at the time it suffocated in tree resin. The same stones also yielded remains of 15 different forms of extinct flowers ranging from 1 to 5 mm in length.

The more we look at the fossil record of the mid-Cretaceous the more we find insects dependent on flower products. In 2019, an extinct wasp, *Prosphex anthophilos*, was described from a deposit in a Myanmar mine. As it died, it vomited eudicot pollen suggesting it had an interior storage bag or crop like a few modern wasps. Let's not forget true flies either, as extinct species of *Buccinatormyia* also appear in Cretaceous amber from Myanmar and Spain. Each fly had a proboscis as long as its body and probably lived on a liquid diet. They carried pollen of cone-bearing plants and most likely drank the pollination drops secreted by exposed, unfertilised seeds. In fact, flies with much elongated mouthparts are older than the Myanmar amber with the fossil of *Archocyrtus kovalevi* dated around 160 million years ago. It came from one of the more typical, compression fossils in sedimentary rock. This extinct member of the hunchback flies (Acroceridae) had a tongue almost twice the length of its body. It probably drank whatever rewards were offered by cone-bearing trees and bushes that presented their droplet oozing ovules in urn-like structures. Today, long-tongue members of hunchback flies in the genus *Psilodera* continue to pollinate native, autumn-flowering *Gladiolus* and *Bulbine* species in South Africa.

I must emphasise that Myanmar amber gives up the remains of moths as well. Most importantly, it yields at least four ancient species of jaw moths (*Sabatinca*). These insects belong to a lineage that goes back even earlier to the Jurassic. Most importantly, jaw moths have only a short proboscis. Instead of a long, coiled, tubular tongue, they have chewing mouthparts and grasping palps like beetles or cockroaches. It's no surprise that jaw moths are often regarded as the most primitive members of the order Lepidoptera. Species of *Sabatinca* are active in tropical forests to this day and are crunching the pollen of flowering trees and the oil containing spores of ferns.

Back in 1999, David Grimaldi, a curator of insects and their fossils at the American Museum of Natural History, wrote a somewhat prophetic review of insects and early flowers of the Cretaceous before the recent discoveries of *Nanjinganthus* and the current flood of descriptions of ancient plants and insects from Myanmar amber. Dr Grimaldi noted that compression fossils made of rock slabs far older than the Cretaceous contained the remains of pollen in the guts and coprolites of insects squashed under layers of sediment. Creatures resembling ice bugs (grylloblattids), sawflies and an extinct group of long-winged and long-tailed insects known as the hypoperlids were consuming the grains of the earliest trees that made seeds but lacked signature carpels or pistils. Combining specimens in rock sediments with some from Cretaceous amber, Grimaldi looked at the veins in the wings of pressed insect fossils from Spain, Siberia, Lebanon, Canada, Brazil, New Jersey and other locations. Wing veins remain a powerful tool for identifying insects to family. Grimaldi showed that wasps, moths and flies (can't say why he was uninterested in beetles) had ancestors in the Cretaceous. In particular, he concentrated on an ancient group of flies in the suborder Brachycera. These ancestors ultimately produced the families of tangle vein flies, hover flies, drone flies, bee flies, and hunchback flies that pollinate true flowers today.

Are there animal pollinators we may exclude from the flowers that bloomed in dinosaur forests and lakes? Yes, we can say that the mid-Cretaceous is far too early to consider the immediate ancestors of bats (Chapter 10) and nectar-feeding birds as early pollinators. It's interesting to note that the geological jury remains out on the origin of wingless mammals that pollinate flowers. Excluding Antarctica, the Southern Hemisphere still contains some plants pollinated by small, often climbing rodents, marsupials, a few lemurs, and some intense little beasts that people misidentify as shrews.

For example, the Cape rock elephant-shrew (*Elephantulus edwardii*) of southern Africa is no climber but it devours insects it finds in wildflowers. These little mammals carry pollen of blooming bulbs like the white pagoda lily (*Whiteheadia bifolia*). When they are thirsty, they drink the flower's nectar and their long snouts are dusted with pollen. As the mammals of the last Age of Dinosaurs were relatively small, one wonders if some ancestors of modern marsupials and placentals climbed into small trees and shrubs to feed. This would account for old lineages of woody plants in which many small flowers are compressed together on branches but their sexual organs are tough, wiry and provide nectar-drinking mammals with ladder rungs. Australian members of the eucalypt (Myrtaceae) and banksia (Proteaceae) families are best known for such flowering branches and some shrubs are dependent on mouse-sized, tree-climbing possums (*Tarsipes* and *Cercartetus*) for pollination.

Therefore, attempting to pin a solitary or ur-pollinator to the first flowers is useless and it doesn't really matter whether they lived in the Jurassic or the later Cretaceous. Insects that ate and/or collected pollen or drank sugary secretions were already at work before any plant enclosed its immature seeds in a pistil bottle. It's far easier to theorise that the first flowers received visits from a variety of unrelated insects. Specialisation probably came later. We simply can't watch the antics of creatures on flowers that went extinct 100 million years ago. As we have no idea how the earliest flowers painted, perfumed or flavoured their pollen or nectar we can't determine whether they first opened their buds during the day or night.

## SHOULD WE RECOGNISE POLLINATION 'SYNDROMES' OR WHEN VERY DIFFERENT PLANTS SHARE THE SAME POLLINATOR?

As Charles Darwin intensified his studies on the floral mechanics of orchids, primroses and many others, it attracted the attention of contemporary naturalists in Europe, Australia, North America and the tropics. These men and women emulated his work while he lived and later generations continued Darwin's investigations into the next century. By the 1960s there was a lot of literature on who pollinated what. This attracted the Norwegian botanist Knut Faegri (1909–2001) and the Dutch botanist Leendert van der Pijl (1903–1990). They reviewed the pollination literature accumulated over a century, made their own observations and came to the same conclusion. Flowers varied because the foraging habits, perceptions and nutritional needs of different pollinators varied. The umwelt of a honeybee didn't quite fit the umwelt of night-flying sphinx moths. The populations of pollinators in every habitat represented another limited resource for each plant species. Plants must compete for the attention of certain animals to be their sperm spreaders, just as any wildflower, tree or shrub competes for sunlight, and their seeds compete for water and available places to sprout. As survival of the fittest really means having lots and lots of generations of offspring it was inevitable that natural selection pushed the evolution of some flowers into intimate relationships with only a few groups of animals at a time. This accounted for the explosive diversification of floral forms over millions of years on different continents and islands.

*The Principles of Pollination Ecology* by Faegri and van der Pijl was published in 1966 and has since enjoyed two more editions (the third appeared in 2013). By comparing hundreds of flowers and their pollinators, the authors came to the conclusion that interrelationships between flowers and certain animals could be inferred if you looked at the flower's physical and chemical traits. Based primarily on floral shape, presentation on a stem (erect, horizontal, nodding), colour and smell, the authors subdivided these relationships into syndromes and it was possible to segregate bat flower syndromes from butterfly flower syndromes, from bird flower syndromes, from bee syndromes, and so on.

The community of pollination ecologists are divided in their opinions. Some insist that these syndromes are too constraining and there are too many exceptions to each category. Others do not agree. They think some botanists and zoologists are ignoring the nuances we could understand much better if we spent more time analysing the chemistries of scent, nectar nutrients and the pigment patterns humans can't see with our own limited senses of vision, smell and taste. Those supporting the Faegri and van der Pijl approach to syndromes argue that their opponents don't go far enough in their field studies.

It is true that some critics of Faegri and van der Pijl don't check the visitors to the flowers they study to see if the animal actually carries the pollen of the same host flower and leaves good grains on the stigmas of the next plant they visit. It's an easy technique and all you need is a microscope and a couple of stains that either colour the walls of all pollen grains or the pollen tubes they sprout and poke down into stigmas. A flower can have lots of visitors but some are mere tourists, nectar thieves or outright vandals. How often does the animal visit the flower? Did it arrive when the stamens released pollen and then returned when the stigma was activated and ready to receive and process incoming grains? Likewise, if a visitor flies to a plant and only visits flowers on the same plant for hours it can't be considered an agent of cross-pollination. Remember, the ability of carpels and pistils to recognise and reject self- and incestuous crosses represents an old and dependable safeguard. Furthermore, pollen sperm, like human sperm, has a limited period of viability once it leaves the stamen where it originated. Do slower animals behaving like gourmands at an all-you-can-eat buffet make good pollinators (Chapter 8)? Field equipment and laboratory techniques have grown increasingly sophisticated so we should be answering these questions.

Don't dismiss me as a traditionalist. There are times when I'm more than happy to agree with the critics. Let's be frank. There is no such thing as ***the*** bird flower or ***the*** fly flower. However, it's obvious that Faegri and van der Pijl didn't believe in a rigid set of absolutes either. Were either of them alive today they would argue that there can't be ***the*** fly flower because there are thousands of species of flower-pollinating flies classified in several unrelated families. In a few chapters, we will learn that mosquitoes and fungus gnats are flower pollinators too, but they do not see or smell the world in the same way.

What we do need to understand next is how and why the flowers of closely related species vary so much that they became pollinated by very different animals? Orchid flowers, in my opinion, make great models to help us understand this common phenomenon. Until then, this chapter began with Robert Herrick (1591–1674). It's fitting he gets the last word.

> *'I saw a fly within a bead*
> *Of amber cleanly buried;*
> *The urn was little, but the room*
> *More rich than Cleopatra's tomb.'*

– The Amber Bead, *Robert Herrick*

# 2
# Getting ugly with the goddess

*Fluellen: '... one Bardolph, if your Majesty know the man. His face is all bubukles and whelks and knobs and flames o' fire; and his lips blow at his nose, and it is like a coal of fire, sometimes blue and sometimes red, but his nose is executed, and his fire's out.'*

– Henry V: *Act 3, Scene 6, William Shakespeare*

IT ALL BEGAN WITH THE VICTORIAN SENTIMENTALITY TOWARDS FLOWERS in general, and orchids in particular. As methods of sea transport and glasshouse culture improved during the 19th century more people in Western Europe began importing more plants from other continents. A modest glasshouse, heated in winter, was a fine investment for the middle classes who could not afford grand estates or hire the army of gardeners needed to grow groves of bamboos, Asian cherries, magnolias, camellias, and rhododendrons arriving from India and China. They were happy to make do with smaller tropical plants and that included a fantastic array of exotic and tropical orchids.

To understand how some pollination systems diversify over time as descendants 'shift' away from the original flower forms made by their ancestors we need to select a lineage of plants to serve as our model system. Preferably, it must be a diverse, but closely related group showing extravagant variations, although sharing the same family tree. Orchids, with over 20,000 species in the family Orchidaceae, are a good choice as they combine the familiar with the exotic. The history of captive orchids in Western civilisation also gives us a good lesson about how we've segregated species based on what we think is beautifully unique.

## DARWIN AND HIS ORCHIDS

Of course, you don't always have to go to the tropics to view orchid diversity. Mild, temperate zones in both hemispheres support less diverse but still choice populations of native orchids. The flowers of wild orchids native to Britain caught the eye of Charles Darwin (1809–1882) while he was a student at Cambridge University. He would collect and preserve a few species on his visit to cold Tierra del Fuego during his voyage on HMS *Beagle*. It's obvious he wanted to know how the organs in an orchid flower worked and why they took such unusual shapes. That meant seeing more of the plants imported from tropical Asia and the Southern Hemisphere. In 1844, he began corresponding with the director of Kew Gardens, JD Hooker (1817–1911), about 'irritability' in orchid flowers. By 1857, he wrote to one of his cousins asking him to view a private collection and observe the flowers of goblin orchids (*Mormodes*) from Central America to see 'which eject the pollen masses when irritated'.

Over at least 20 years, Darwin combined his observations and dissections of species growing wild from the fields and woodlands around his country home with experimenting on treasures loaned from Kew Gardens and from passionate collectors, such as James Bateman (1811–1897). His investigations led to the 1862 publication of a book that appeared only three years after the explosive *On the Origin of Species.* It was entitled, *On the Various Contrivances by Which British and Foreign Orchids Are Fertilised by Insects, and on the Good Effects of Intercrossing.* It addressed the flowers of 70 orchid species. A second edition published in 1877, under a thankfully shorter title, added another 40 species from all over the world, including Australia.

The second edition is still worth reading because the author has something intriguing to say about the mechanics of so many flowers.Most of the work in the first edition requiring direct observation under magnification, dissection and experimentation was performed primarily by Darwin alone and it was not possible for him to get it right every time. His slow expansion into writing a second edition allowed him to correct his own errors as, during the 15 intervening years, his first edition attracted botanists in other countries who wanted to correspond with him.

In the first edition, Darwin examined and dissected flowers of slipper orchids. In his day, any tropical Asian plant producing a flower with a petal that resembled a shoe or bucket was placed in the genus *Cypripedium*. Today we know these Asian orchids as *Paphiopedilum* species and their hybrid descendants remain very popular with hobbyists. To Darwin, they were formidable flowers, perhaps a little intimidating.

All the other orchid flowers he examined had only one pollen-making stamen, except paphiopedilums which have two and are separated from each other by a central, large, coloured and sculptured organ known as the staminodium. The staminodium forms an inflexible shield over the thick pollen-receptive female organ (stigma) concealed underneath. The lip petal (labellum) of the same flower is an inflated shoe-shaped structure encircling the stigma and hiding it from view even further. As the flower bud opens the lip curves and puffs out, and also makes three openings along its margins that allow an insect to enter the shoe, pass under the stigma and exit unharmed. On its way out the insect contacts one of the stamens picking up the flower's gluey pollen before it can fly to freedom. The largest opening is a long, wide vertical slit along the top of the shoe-shaped lip petal. An additional two much smaller openings are located in the rear, on either side of the staminodium. Each of these smaller portals encircle one of the two pollen-producing stamens. As the staminodium of a paphiopedilum flower is shiny and coloured, compared to its greenish-brownish lip petal, Darwin logically assumed that the pollinator was drawn to it first. From there, the insect could enter the flower by one of the two rear holes helping itself to pollen provided by either exposed stamens. Crawling down this rear hole it entered into the hollow shoe, walked under the stigma and then let itself out by flying up through the wider vertical slit on the shoe's upper surface. So Darwin wrote in the first edition of his book and he was ... wrong!

Darwin knew of his mistake at least a year after the publication of his first edition. He'd been corresponding with Asa Gray (1810–1888), the first professor of botany at Harvard University, who had been studying American slipper orchids (*Cypripedium*) with enthusiasm after reading Darwin's first edition. Gray proved more adept at solving floral puzzles. In a letter to the South

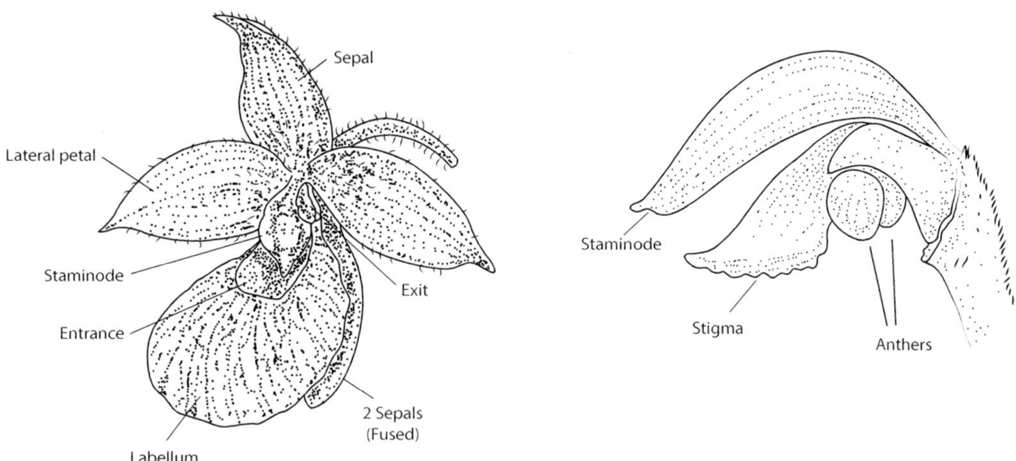


Diagram of a typical slipper orchid (subfamily Cypripedioideae) showing the sac-like lip petal and the sterile stamen that has become a shield-like staminodium. Illustration by Envisage Information Technology.

African naturalist Roland Trimen (1840–1916) in 1863, Darwin wrote, 'By the bye, I believe I have blundered on *Cypripedium* [*Paphiopedilum*]: Asa Gray suggested that small insects enter by the toe [vertical slit] & crawl out by the lateral windows [rear holes].– I put in a small bee & it did so & came out with its back smeared with pollen ...'.

Scientists should learn a lesson from Darwin's humility. He read the description of a trusted colleague, admitted his blunder and set it right with an experiment he devised, and added it to the second edition. What is of greater importance is that the architecture of flowers of the more than 175 species we know as slipper orchids forces each visiting insect to take the exact same route. The visitor enters through the large vertical slit or canyon on the lip petal. If it is already carrying pollen on its back the pollen will be scraped off as it passes under the brushy stigma concealed inside the shoe *under* the staminodium. To escape, the insect must head for the base of the lip petal and crawl up one of two canals to reach one of the escape hatches. As it struggles to push itself out of a rather tight hole, it brushes against an anther, which smears fresh pollen on its back. It doesn't matter whether the flower belongs to a *Cypripedium, Paphiopedilum*, *Selenipedium* or *Phragmipedium.* If the insect is a true pollinator of the flower there's only one legitimate way in and a second, but opposite way out that will leave any pollen it's already carrying on the stigma and then sends the same insect out into the world wearing a fresh glob of pollen.

The scientific names of the first three genera belong to goddesses with pretty feet (pedium) or wearing fancy shoes (pedilum). The Cyprian one and the Paphian one refer to the love goddess, Aphrodite, her loyalty to the island of Cyprus and its port of Paphos after she rose from the sea. Selene, was a Grecian goddess of the moon and there's one thing we know about goddesses, in general. Their myths insist they appear to us mortals in more than one aspect. Hinduism portrays the warlike and hideous Kali killing demons versus her benign and beautiful avatars as Durga and Parvati. Mythographer Robert Graves (1895–1985) believed all female deities of the Mediterranean were triple goddesses appearing as a maiden, then a mother, and

Follow the arrow. This is the pollinator's pathway when visiting the inflated lip petal of the Chinese *Cypripedium wardii*. The stigma should scrape off pollen as the insect passes under and either anther will smear more pollen onto the same spot as it escapes via one of the two exits. Illustration by Envisage Information Technology. Right, while the lip petal *C. wardii* is decorated and obvious, the white staminodium fades into the background. Photo by Jidong Ya.

then a crone. This provides use with an unnatural but useful allusion to natural variation in flower presentation.

## DRAWBACKS TO DECEITFUL POLLINATION AND ITS IMPLICATIONS FOR CONSERVATION

No species of slipper orchid enjoys a perfect pollination system, regardless of the shape, colour or scent of its flowers. Darwin ruled out the perfection of adaptations by what the proper Victorians called providence. Indeed, research indicates that pollination is rarely a sure thing in slipper orchids. Experimenting with live flies or small bees, Darwin observed that some insects were too slow or stupid to follow the one-way path or too weak to squeeze through the rear exits. Based on my own fieldwork, I've found my share of corpses inside these fashionable shoes. The victims probably expired from a combination of starvation and/or exposure. Some visitors are too small to contact the stigma or the anthers on the way out. The tiniest escape quickly through the rear portals without picking up any pollen at all. Big visitors may find the shoe too tight a fit. They fret and roll about in the petal until they bounce out of the flower by the original entrance and the plant gains nothing. In one case, I watched a chunky leafcutter bee (*Megachile*) use her powerful jaws to rip her way out of the chamber of a mountain lady's slipper (*Cypripedium montanum*). She escaped after slicing a hole through the floor and the lip petal deflated within a few hours. Of even greater importance, after an uncomfortable experience it appears that some insects refuse to repeat the process and won't visit a second flower even though they are wearing pollen on their backs or faces. Without a visit to a second flower on a second plant cross-pollination can't occur.

Why should any insect try a second flower unless they are really naïve, in the first place? Most slipper orchids are traps. There's nothing to eat on or inside the flower. Pregnant hover flies (Syrphidae) mistake the warts and bumps on the staminodium of several *Paphiopedilum* species for colonies of live aphids. Their maggots mature by devouring such green or black plant lice (see below), so when the pregnant fly attempts to lay her eggs on staminodium warts she slips on the polished surface and falls through the big slit into the chamber of the lip petal. This, one presumes, becomes frustrating after one or two attempts and flies stop visiting when they don't make any progress. The flower isn't built to pollinate itself. If the fly fails to enter more than one bucket on more than one plant, the cheating orchids can't make seeds because there aren't enough suckers willing to repeat the humiliating process twice.

As there is no such thing as a perfect mimic the fertility of slipper orchids varies enormously from species to species, from site to site, and from season to season. At Grande Ronde Overlook Wildflower Institute Serving Ecological Restoration (GROWISER) , a wildflower reserve in Oregon, USA, the populations of mountain lady's slipper were so well visited by small bees with poor memories that 75–85% of the flowers developed into fruits over two seasons. The same couldn't be said for eastern populations of the American pink moccasin flower (*Cypripedium acaule*) regardless of site. In a year of low pollinator activity, a little more than 1% of flowers set fruit. In a good year, it's closer to 25%. Likewise, in one season scientists monitored hover flies visiting populations of two tropical Asian slippers. While 90% of the flowers in a population of *Paphiopedilum callosum* were pollinated no fruits developed in the similarly distributed *P. charlesworthii.* Fly visits to the second species were far too few.

Such boom and bust cycles among slipper species are important to conservationists trying to maintain biodiversity in threatened areas. My Chinese colleagues insist that once-extensive populations of hundreds of slipper orchids in the mountains of Yunnan have shrunk within the past 15 years. In general, slipper orchids live only in old, stable habitats within monsoon climates that weather porous rocks into chemically suitable soils. It's typical for individual slipper orchids to take years to grow to a respectable size accumulating enough resources to make only a few flowers every year. The positive side is that some slipper orchids live for decades, like elephants, and, like elephants, don't have to reproduce offspring every year to maintain a population. As their flowers are bisexual, one plant may make a valuable contribution in one season by fathering seeds even though it fails to become fertilised. Unfortunately, receding glaciers, insufficient rains, encroaching urbanisation, agricultural expansion and poaching may be changing this status quo in China, just as it has in North America.

There's no guarantee that those few fruits will mature to release their seeds either. My research partner, Retha Edens-Meier, found that fruits of the Kentucky lady's slipper (*Cypripedium kentuckiense*) started developing in early summer but didn't release their seeds until the following spring. That's a long time to leave a juicy stalk exposed to deer and bears. We also worked together on the showy or queen's lady's slipper (*C. reginae*) in southern Missouri. When Retha returned to our favourite site, a couple of months after the orchids' flowered, she was pleased to see that a few ovaries were swelling only to feel aghast when a large, obese caterpillar in the family of looper moths (Geometridae) gnawed through a fruit wall. One Chinese colleague once told me about a site in a reserve where the slipper orchids never set seed because they were devoured by domesticated yaks. 'Can't you fence them out of a protected reserve?' I asked. His reply was curt. 'Yaks are very valuable in China'.

We can't expect to protect every endangered species by growing them on in pots. In fact, some *Cypripedium* species refuse to thrive in captivity. Understanding how slipper orchids make more slipper orchids is a priority as the reproductive health of their remaining populations often tells us more about the density and diversity in the forests, mountain slopes and stream banks where the last plants grow and their pollinators fly. I helped work on these problems in America and China with brilliant and concerned colleagues since the early years of the 21st century and I'm still learning.

## 'MAIDEN' VERSUS 'MOTHER' VERSUS 'CRONE' POLLINATION IN NORTH AMERICA

What we've all learned is that no slipper orchid flower can appeal to every potential pollinator living in the same ecosystem. One floral size, colour or scent doesn't fit all. Different species of slipper orchids may share the same habitat and bloom at the same time but are often pollinated by very different visitors. We've made some progress with the 50–60 *Cypripedium* species distributed in North America and Eurasia. These are the orchids tourists prefer and want us to protect because they produce some of the largest flowers. Their lip petals range in size. Some are only the circumference of a hazelnut while others are larger than a tea cup. Their lip petals may be pure white or bear rich yellows, pink or shades of purple. Some smell like a bag of lollies or an expensive perfume. These species are almost always pollinated by bees, but different flowers interact successfully with different bees from species to species.

Think of the yellow lady's slipper (*C. parviflorum*) and the mountain lady's slipper we examined previously as the maiden goddesses of Robert Graves. In America, they are bee generalists. As their lip petals are hazelnut in size they are most successful when they lure smaller bees 5–9 mm in length and only 2–3 mm at their widest points. Most bees are of similar sizes (the commercial honeybee is an atypical, big female) and we recorded many species, belonging to three different bee families exiting the flowers of our yellow and mountain maidens wearing swatches of pollen on their backs. They visited the flowers frequently, especially when the plants stood in forest light gaps, for an hour or two, at discrete times during the day. We noticed that within 15–20 min after they basked in the light and warmth of a woodland light gap the flowers released their pleasant aromas.

Moving on to the larger mothers, we return to the most aptly named showy lady's slipper of Canada and North America with its big, pink or pink blotched and speckled shoe against a white background. As expected, it attracts larger bees such as leafcutters, chimney bees (*Anthophora*) and even introduced honeybees (*Apis mellifera*). The same flower is so attractive it also lures the tiny miner (*Lasioglossum*) and sweat (*Augochlora*) bees. This is a waste of time and effort for the flower. These little bees enter the comparatively cavernous lip petal and exit in less than a minute without contacting the stigma or pollen as if they were on vacation enjoying the big slide at a water park. Unfortunately, those larger bees, while effective pollinators were never frequent visitors. We found that over several seasons only 4–33% of showy lady's slippers set fruit in our Missouri sites.

Let's take it up one more notch and return to the American pink lady (*C. acaule*) and add the Tibetan slipper (*C. tibeticum*). Both species have large, deep purple lip petals. They attract only big, broad bumblebees. In fact, the American pink slipper blooms at a time of year in which, if it's visited at all, it is pollinated exclusively by queen bumblebees that hibernated over winter. The Tibetan flower is visited by a combination of queens and workers, but its success rate of fruit set is only 9–25%. One wonders if a big bee is a little smarter and harder to dupe.

Well, I promised you crones, didn't I? They don't receive as much attention from photographers or artists as they lack large, luscious and brilliantly coloured flowers and they come with some unpleasant looking pollinators. While maiden and mother *Cypripedium* species vary in their abilities to exploit bees representing different sizes, castes and foraging habits the crone slippers eschew bees entirely. Pollinator shifts have obviously occurred in *Cypripedium.* Thus, we might ask, what happened over time? Did the ancestors of some slipper orchids colonise places where bees were uncommon or they bloomed at odd times when bees were infrequent? Did these ancestors lose bees to the flowers of more attractive plants that offered good and consistent rewards? In either case, natural selection favoured those slipper orchids that attracted atypical pollinators. Once again, survival of the fittest is measured only by the fertility of the generations you leave after you die.

There's something to be said for the theory that some crones fit best in places bees dislike, My first experience with a less than glamorous slipper orchid was the clustered lady's slipper (*C. fasciculatum*) in populations in Oregon, Colorado and Idaho with Dr Nan Vance and her team from the USDA Forest Service. These plants shrivel up in hot exposed places growing in the dense, cool shade of old Douglas firs. The forest floor is moist and gloomy while this orchid blooms and most bees dislike foraging in excessive shade. A stalk of a clustered lady's slipper bears one to four drab, brownish or greenish, nodding flowers. The lateral petals and dorsal sepal are so long and droopy they often obscure the plump little gumdrop of a lip petal.

Our job was to determine how often the flowers were pollinated, and the conversion rate of flowers into seed-filled fruits comparing results over several seasons at several sites. This we did, but the pollinator remained elusive. However, something was enthusiastic about visiting these small and rather morose flowers of dark forests as, given the year and site, 18–70% of the flowers set fruit.

We obviously weren't looking in the right way or at the right time. Carol Ferguson and Kathleen Donham, then of Southern Oregon University, spent four years monitoring populations of the clustered lady's slipper as their blooming peaked in May. Yes, they did find insects that entered the slit on the lip and exited via the rear portals carrying swabs of pollen on their backs, but the only pollinators were tiny, female wasps less than 5 mm in length. These diapriid wasps, in the genus *Cinetus,* were more than happy to remain in the shade. Why? They have a parasitic lifecycle laying their eggs in the bodies of maggots of fungus gnats swarming in the humus of the forest's floor. The maggots eat the fungal threads that grow in the decaying compost of needles under the trees. Perhaps some gnats also lay their eggs on real mushroom caps or stalks. Perhaps the wasps mistake the orchid's shoe for the parasols and toadstools that serve as maggot nurseries. Some people smelled the flowers and insisted they detected a mushroom-like aroma.

It's an unusual chain in the web of forest ecology as it's not based exclusively on who eats who. The tree drops needles and twigs that feed and house sprouting fungus spores. Their mushroom threads and fleshy bodies are eaten by maggots. Tiny wasps turn the maggots into living kitchens and cradles for wasp larvae. The orchid exploits maternal wasps, not pregnant gnats. As usual, the lip petal of the orchid offers no nectar or anything to consume, so some sort of mimicry is occurring. That may have a lot to do with the flower's fragrance, but no one has analysed it yet. The natural irony here is that in chapter 5 you will read that there are hundreds of orchid species in Australia dependent entirely on male or female fungus gnats for pollination. Indeed, Chapter 7 also shows that many other Australian orchids are also pollinated by parasitic wasps. These are always male wasps and belong to species that have no interest at all in the maggots of fungus gnats.

## CHINESE CRONES

Additional studies on slipper orchids in China tell us more about how floral scents influence the behaviour of insect pollinators. There is one crone found from Tibet to northern Vietnam that balances deceit with treats, based on field and laboratory work by Yung-I Lee and his team. To study this unusual relationship, they travelled to Yunnan from their base at the National Museum of Natural Science in Taiwan. The flower of *Cypripedium subtropicum* is mostly chocolate and tan. The side lobes of its lip petal stand out against the homely flower because they bear irregular tufts of white hair against a cocoa-coloured background. The orchid looks like it's suffering from a bad case of dandruff. Under a powerful scanning electron microscope each hair looks like a string of conical beads and it's suspected that the six cells that make up each hair chain are packed with nutrients. The tufts are visited by at least 10 species of hover flies that eat the hairs. Chemical analyses showed that these hairs contained some sugars, but were unusually rich in amino acids, the building blocks of proteins.

The food in these hairs must be very important to these insects. When Dr Yung-I and his crew 'shaved off' the white tufts on the flowers the flies stopped visiting. The gluttony of hair-gobbling flies makes them careless. They crawl over the lip looking for more snacks falling through the slit into the shoe's chamber. If they are already carrying orchid pollen on their backs, it will be received by the hanging stigma as they head for the exit canals. When they squeeze out of the portals, more pollen will be stuck to their backs. At least six fly species were observed carrying the orchid's pollen. In a good year ~18% of the flowers set fruit after flies visit more than one flower.

Where does the deceit come in? At least 70% of the hover flies found visiting this orchid were females. As noted earlier, hover fly females lay eggs amongst aphid colonies. When their larvae hatch, they ravage the defenceless flock. However, female hover flies also drink the honeydew that aphids excrete. While hover flies have good colour vision they also locate aphid colonies by the scents their prey secretes when they are disturbed and alarmed. Aphids in distress warn their sisters and offspring with odour molecules, such as (E)-beta-farnesene and beta-pinene.

The flowers of *C. subtropicum* have a pleasant fruity odour to the human nose combining at least 10 different scent notes. Some of these components are so familiar they remind our noses of citrus skins and ripe mangoes. As predicted, two scent notes produced by the orchid also mimic the 'warning scents' of an aphid colony under attack and the floral fragrance is rich in β-pinene and (E)-beta-farnesene. Dr Yung-I set out sticky traps inside small, open bottles that were baited with either bits of real flowers, a synthetic perfume made of all the known scent components, or each of the pure, distilled essences of the two main warning scents. As predicted, the local hover flies preferred the real flower baits. However, some entered the bottles containing the synthetic perfume, while others were trapped in the bottle of pure (E)-beta-farnesene or the bottle of pure beta-pinene. In contrast, the same flies showed complete disinterest in additional bottles spiked with more common chemicals human noses associate with ripe fruits.

Pure deceit rules in ongoing studies of another slipper orchid in temperate China. This flower really does appear to mimic fruit, but its colours and scents differ from those expressed by *C. subtropicum*. Professor Luo Yi-bo asked for my help to sort out the story of *Cypripedium bardolphianum*. This is another tiny-flowered slipper orchid, and grows on mossy and rocky slopes in the southern provinces of Sichuan, Yunnan and Gansu in China. Once again, it's a small plant and makes a pouch-like lip only 8 mm in length. That lip bears so many tiny, reddish bumps that the two botanists who described it as a new species in 1916 had a sense of humour and insisted it matched the face of Bardolph, one of Shakespeare's ugliest characters. Two reddish-brown lateral petals clasp the lip petal as if it needed a chin strap. The lip is also partially hidden under its own dull-coloured cap of a dorsal sepal.

Yi-bo and his students found a population in Sichuan and compared fruit production in 2004 and 2005. While at least 500 plants bloomed each year, the development of flowers into seed-filled capsules was never higher than 13%. It took 143 h of observations to find and collect the infrequent pollinators and they were miniscule. In the case of Bardolph's slipper, the only insects to carry the pollen packets were male and female fruit flies (*Drosophila*) of an unknown species.

What would make a fruit fly visit a flower instead of a fruit? The obvious answer is a flower that looked and smelled like an overripe berry or fallen fig. Yi-bo's team analysed the slight but musty odour of fresh flowers and it produced three recognisable peaks. There was a pleasant fruity note of ethyl acetate (70%). High concentrations of ethyl acetate are insect-killing poisons, but at very low concentrations the smell reminds you of ripe pears and fails to harm the tiniest creatures. This was followed by the vinegar-like aroma of acid butyl ester (23%). A third molecule, known as 1-Butanol (6%) contributed to the illusion of alcoholic fermentation. Since we don't know how well fruit flies see their world we are left to wonder if the shiny red bumps on the lip petal (Bardolph's bubukles) mimic juice droplets or the exposed seeds of a ruptured berry or fig.

Yi-bo was generous. He let me join a team of scientists using molecular genetics to construct a family tree of all *Cypripedium* species. We collected, pressed and dried leaves of known species and sent them to laboratories comparing the genes in their cell nuclei and photosynthesis factories, known as chloroplasts. We learned that China was the real centre for the march of the crones. The 'sister' of my American clustered lady's slipper was an equally small, droopy wildflower of mountain slopes, restricted to Sichuan, known as *C. palangshanense*. Its lip petal is deep, shiny and purplish-red like coagulating blood. We still don't know what pollinates this flower. Bardolph's slipper has two additional Chinese sisters, both homely and small. They have yellowish-green flowers with brown checks and speckles. We don't know their pollinators either, but it's obvious they must attract small insects unimpressed by vivid pigments.

There was something more about the family tree that proved to be very Chinese and very unusual. Six species formed their own branch in between Bardolph and her crone sisters and a branch featuring the beautiful queen lady's slipper and her two equally lovely relatives. The six new crones were confined to their own twig. When they bloom, their flower stalks are so short they hardly peek above the forest's leaf litter. Their green leaves are spotted and hairy. Some of them offered flowers resembling fuzzy, brownish toilet bowls. Not all the species made tiny flowers either. On a field trip to the Yulong Mountain in Yunnan province I was introduced to *C. lichiangense.* The lip was 7 cm in length and 5–6 cm in width. It was flecked with dozens of liver spots. Touching the shoe and lateral petals for the first time gave me the creeps. The texture reminded me of the cartilage in my own ear. We found a dead fly in the shoe but it wasn't carrying pollen. The mystery of this bloated, woodland bedpan was solved by CC Zheng, a student of Yi-bo's, with access to a large population in bordering Sichuan province.

Why did the thickness and texture of the flower remind me of ears? Because it was built to last. Once the bud opened, the blossom was expected to live at least 25 days and the population of 1,000 plants stayed in bloom for over 2 months. They were unusually patient plants waiting for the right dupe to enter their shoes. Zheng insisted the flowers smelled like rotting vegetation. When flies approached the lip they flew in with a sort of zig-zag pattern indicating they were homing in on its scent. Zheng spent days, and some evenings watching those flowers, totalling more than 340 h.

While flies from several families, some beetles and *Cinetus* wasps, visited this strange flower, only one insect species entered by way of the front slit on the lip petal and exited via one of its two rear holes. They were all female hover flies of *Ferdinandea cuprea.* Out of 18 visits, four actually exited the flower carrying pollen globs on their backs, and they were slow. It took at least one fly 1 h 46 min from the time of her entrance into the lip to her escape. No wonder the flower was built to outlast its pollinator. But these insects did something before they left the flower that no other lady's slipper pollinator had ever been observed to do. Some of these flies left their eggs in the lip or hanging from other floral organs. This crone was pretending to be the natural brood site of the fly. Unlike the hover flies mentioned above that lay their eggs in aphid colonies, the maggots of *F. cuprea* are vegans either preferring sites of long-dead vegetation rotting down into humus or wet, diseased roots leaking sap. The long lifespan of the individual flowers of this lady's slipper strongly suggests that local supplies of naïve flies, so stupid they spend hours cross-pollinating two flowers, are rare. The fruit set of over 500 flowers were monitored over two seasons. The conversion rate of crone flowers into fruits was only 6–7%. If you're wondering, the answer is yes, fly maggots that hatched in the flower starved to death.

Just as the maiden and mother species of lady's slippers vary in size, colours and scent appealing to bees of different sizes and foraging behaviours, so do the crone species of China, but their attractants appeal to insects with very different priorities. The sex life of the tiny Bardolph's slipper operates with visits from equally diminutive male and female fruit flies. The big weird sister *C. lichiangense* relies on bigger, chunkier and pregnant hover flies. The remaining species of what I've called crone slippers will be of interest to Chinese scientists for many years provided they can find big populations to study.

For example, not all species of hover flies are content with the aphid honeydew or even the nectar and pollen offered by more honest flowers. Some eat the spores and/or lay their eggs on plants dying of diseases caused by fungi. This may explain some of the species found on the same family twig as *C. lichiangense.* They share the same spotted flowers and hairy foliage. According to Dr Zong-Xin Ren of the Kunming Institute of Botany, *C. fargesii* may be a drama queen pretending to die of a morbid infection. This crone is pollinated by the hover fly, *Cheilosia lucida.* These insects visit diseased foliage and are known to eat the spores of fungi emerging from dying leaves.

Now you can see that the orchids offer a window into the evolution of relationships between unlikely organisms. Their flowers make us reconsider what we think we know about shifts in the evolution of flowers. Let's stay with them and their pollinators, but increase the survival stakes. In the next chapter, we meet a most unobtrusive orchid dependent on one of mankind's oldest enemies.

# 3

# One hundred years of mosquitoes and orchids

*Queer, how you stalk and prowl the air*
*In circles and evasions enveloping me,*
*Ghoul on wings*
*Winged Victory.*

– The Mosquito, *DH Lawrence*

MOSQUITOES NEED A SOURCE OF ENERGY BEFORE THEY TORTURE US. LIKE ALL true flies, a mosquito has only two wings. They must buzz steadily and rapidly, like an outboard motor, pushing the insect's body through the thick atmosphere. On hot, summer nights this explains the sound of the 'hateful bugle in my ear', to quote DH Lawrence. The smaller the body, the more it must struggle against the dense air around it and that takes a lot of energy. So, where do female mosquitoes find enough fuel to fly, as the blood they harvest from us is for provisioning their eggs? From the sugar made by plants, which is quick and easy to drink when secreted by flowers as diluted nectar.

Observations of mosquitoes 'sucking honey' from flowers began to appear in science journals during the first two decades of the 20th century. In 1907, the *Journal of the New York Entomological Society* published a short paper by Frederick Knab (1865–1918). He reviewed the 19th century reports of Europeans who observed *Aedes* and *Culex* mosquitoes visiting the flowers of catbrier (*Smilax*), native hydrangeas and the tiny, congested florets that make up the yellow fingers of goldenrod (*Solidago*) and other members of the daisy family (Asteraceae). Knab made his own observations in Saskatchewan and found male and female mosquitoes frequenting the flower clusters or catkins of bushy willows (*Salix*). He never insisted that these bloodsuckers pollinated the flowers they visited for drinks.

By 1913 though, there was another surprise published in the American journal *Science*. Ecologist Ada K Dietz (1888–1981) was studying plants in Michigan bogs in 1912 when she noticed that local mosquitoes wore discrete masses of yellow pollen on their heads. She told Dr John Smith Dexter (1885–1928) of Columbia University, who was visiting the same field station owned by the University of Michigan. Dr Dexter visited the bog and caught half a dozen mosquitoes, all females, with pollen clusters on their heads. The only plants that compress all their pollen grains into discrete wads (pollinia) attach them to sterile stalks then glue those stalks to pollinators belong either to the orchid or milkweed (Apocynaceae) families. The only wildflowers making sticky pollen units (pollinaria) in the bog that day were blunt-leaved orchids (*Platanthera obtusata*). While Dexter was convinced that the mosquito was the orchid's pollinator, he really didn't follow up his observations with experiments and he didn't examine the flowers to see if any insect had left a gift of pollen on the wet and receptive tip (stigma) of the female organ in the orchid's flower. He also didn't identify the mozzies to species.

These Michigan mosquitoes repeatedly withdrew the complicated pollinarium from the flower. This begs the question: Couldn't they fly to a second flower on a second plant on the same day effecting cross-pollination and ending in a new generation of seeds? As we know from Chapter 2, Charles Darwin published a book on the mechanics or contrivances orchid flowers employed to exploit insects as pollinaria taxis in 1862. He received such a positive response to the first edition that he wrote a revised edition in 1877 containing more of his own research. In this edition he also included the observations of colleagues on other continents. Darwin watched butterflies and moths visit flowers of the pyramidal orchid (*Anacamptis pyramidalis*). Each beautiful insect wore the sticky plug (viscidium) of a flower's pollinarium attached to the base of its coiling tongue. He caught bumblebees in the act of pollinating a lady's tresses (*Spiranthes autumnalis*). Darwin died in 1882. Who would do the field and laboratory work to see if Dexter was right or wrong?

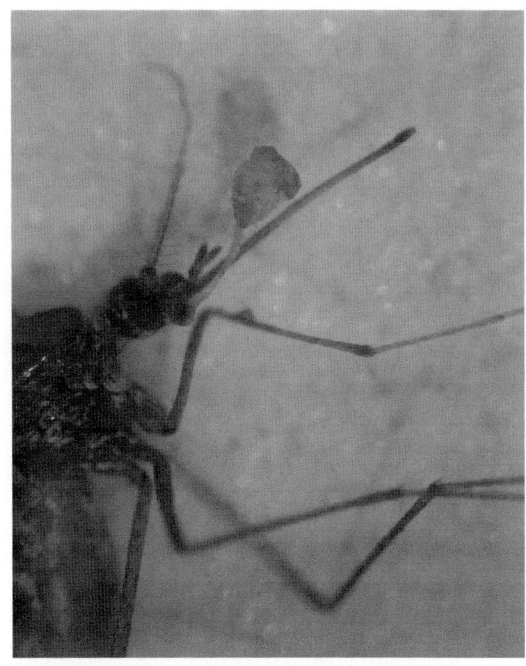

A mosquito carries one pollinarium (1 pollinium + 1 caudicle stalk + 1 viscidium plug) of the blunt-leaved orchid (*Platanthera obtusata*). Photo by Chloe Lahondere.

## PORTRAIT OF AN IGNORED ORCHID

The blunt-leaved orchid can afford to wait patiently for a new generation of scientists. It's a surprisingly successful and widespread plant when you realise it belongs to a largely tropical family. Rooted in peat and mud its flowering stem rarely stands higher than your shin. This is an orchid that doesn't mind a long, frozen winter or flooded and acidic soils, unlike most of its relatives. It blooms in mid-summer from Scandinavia to Siberia through Korea then it appears in Alaska before springing up on either side of the border between Canada and the United States. It may even be found further south. Where it grows in New Mexico, it survives the colder elevations in the Rocky Mountains. Some taxonomists divide it into two subspecies. The North American populations are classified as subspecies *obtusata,* while the rarer *oligantha* is Eurasian.

Defining relationships between the blunt-leaved orchid and the mosquito took additional decades one suspects, because the flowers didn't inspire much awe or professional interest. Who wants to spend the summer in danger of vanishing in a bog? Furthermore, ecologists lost interest

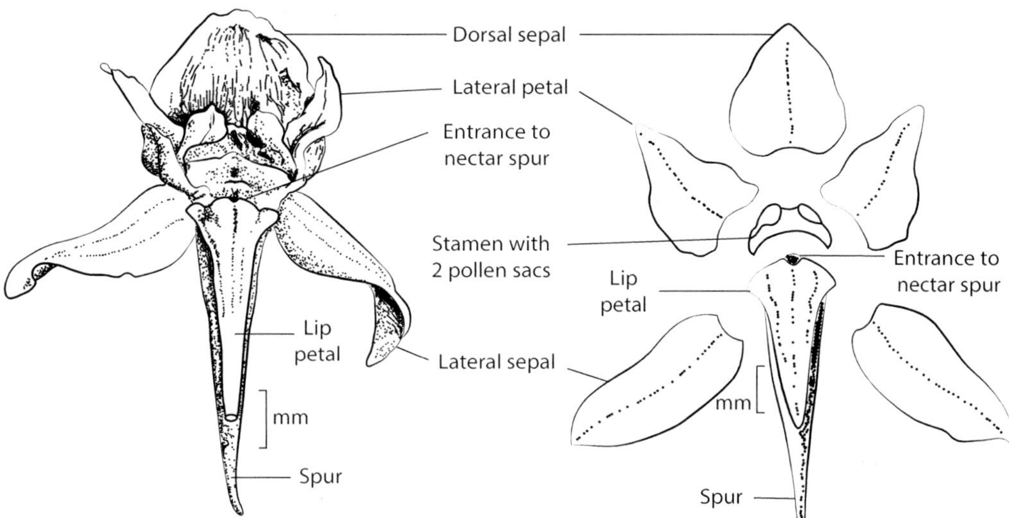


Diagram of the flower of the blunt-leaved orchid. Adapted from Welby R. Smith's *Native Orchids of Minnesota*, 2012.

in pollination field studies during much of the first half of the 20th century. A few gifted and farsighted botanists and amateur naturalists remained focused on the topic, but most biologists weren't terribly interested as they already had books by Darwin and several German botanists written in the previous century. Yes flowers need insects for cross-pollination and the flower pays the pollinator with something edible. So, what else is new? The flowers of a blunt-leaved orchid weren't about to provoke general excitement. You wouldn't pick them and wear them as a corsage. They are tiny and coloured an unexciting yellowish-green with white flecks. However, place one of these flowers under a hand lens and you must agree it has everything you'd expect to see in a larger, gaudier *Phalaenopsis* or *Cymbidium* hybrid from an expensive nursery.

As in all orchids, the blunt-leaved flower is made of three sepals and three petals. As usual, one of the petals is shaped differently from the other but it doesn't form the inflated sac we saw in the slipper orchids in Chapter 2. Instead, the lip petal is a narrow strap. As in some other orchids, the lip petal also tapers and elongates at its base forming a hollow spur. As nectar is stored within the spur's tip, the flower's pollinator must use its long and narrow mouthparts to reach it. The sexual organs opposite the labellum have also fused together into a much simpler column than the one examined in Chapter 2. The column of the blunt-leaved flower lacks the exaggerated staminodium of the slipper orchids. It has only one fertile, pollen-making stamen attached to the back of the solitary, seed-making pistil. One part of the female stigma makes the two sticky, adhesive disks called viscidia. That's the big difference between the blunt-leaved flower and its closest relations compared to most of the other species in the orchid family. In most orchid flowers, there is only one disk or glob of glue and all the pollen balls (pollinia) in the anther becomes attached to it once it's plugged onto the body of the pollinator. In a *Platanthera* flower, each pollen ball (pollinium) has its own stalk, which connects to its own discrete and sticky disk (viscidium). There is one stalk and one viscidium for the pollen ball on the left side of the anthers and one stalk and viscidium for the right. To release both pollinia each *Platanthera* flower must be visited twice, preferably by two helpful insects. That's the basic architecture and mechanics. It doesn't matter whether the *Platanthera* flower belongs to a species pollinated primarily by a moth, butterfly, bee ... or a mosquito.

## WHO CONFIRMED MOSQUITO POLLINATION?

There was only one more brief and early attempt to look at the mosquito and the orchid before the second half of the 20th century. In 1930, Dr Hugh Raup (1901–1995) of Harvard University published a paragraph in *Rhodora*, a New England-based journal of botany. Raup had been to north-western Canada and had collected mosquitoes wearing the blunt-leaved orchid's pollinia. This time there were two convincing photos of a pinned insect wearing whole pollinaria. In fact, one specimen wore three pollinia balls on her head indicating she'd visited at least three flowers before she was caught and killed. Unfortunately, once again, Raup failed to identify the mosquito to species.

From the late 1940s to 1950, several entomologists surveyed the diversity and population explosions of various biting flies in Canada. Of course, mosquitoes wearing the blunt-leaved pollinia turned up and, in some seasons, at least 335 mozzies were collected wearing at least one pollinarium. All the mosquito species bearing orchid pollen belonged to the genus *Aedes*. That was an important discovery, as we will see later, that might have a future impact on public health to protect people from insect-borne diseases.

A real confirmation of Dexter's prediction arrived finally in the late 1960s. Wisconsin's populations of blunt-leaved orchids were receiving the attention of a talented, young botanist named Leonard B Thien mentioned in the Introduction.

Len had two things in his favour. First, he became interested in orchid biology while still a Masters student in Saint Louis. The National Science Foundation funded his position as an assistant to Calaway Dodson (1928–2020), then curator of orchids at the Missouri Botanical Garden. It looks like Len and his mentor much enjoyed a successful orchid-collecting expedition to Ecuador. Dodson was so pleased with his assistant's help that he named several new species after him; *Benzingia thienii, Encyclia thienii*, and *Masdevallia thienii*. One glorious parrot-beaked species turned out to be a natural and recurrent hybrid and is now known as *Stanhophea x thienii.*

Second, over a research career lasting almost 40 years Len proved he was familiar with innovative ways to study aspects of a plant's life that were not obvious to the naked eye. For example, when he became interested in the blunt-leaved orchid his wife Lorraine was worried. Those Wisconsin bogs are extensive and treacherous. Len worked alone at that time and if you fell through one of those floating mats of sphagnum moss, no one would see you again. Fortunately, Len was not the sort to chase tiny insects over untrustworthy terrain waving a butterfly net. In 1967, the *Journal of Medical Entomology* published a paper on how to make mosquito traps using dry ice as bait. After all, we betray our presence to mosquitoes by the carbon dioxide fumes we exhale. Len set up the traps in his bog sites and soon amassed a collection of 462 mosquitoes representing five species. Thirty-eight mosquitoes wore orchid pollinia, but they belonged to only two species: *Aedes communis* and *A. canadensis.* Neither of these species were collected by the Canadian team in Manitoba two decades earlier.

To better watch interactions between such tiny insects and small flowers Len made glass cages, released living mosquitoes he'd captured into them and placed the cages over orchids in bloom. He recorded activities with a camera. Once the mosquitoes grew accustomed to their prison, they began to explore the flowers. The lip petal was too narrow and short for the gangly legged mozzie to use as a perch. Instead, the claws on four of her six legs gripped the two, opposite, wing-like, lateral petals located on either side of the lip. Now she could force her proboscis down the narrow canal made by the lip petal and the column. Of course, she was aiming for the droplet of nectar collecting at the tip of the spur, but she had to push as hard as she could smooshing her face into one of the two sticky viscidium discs. Len found that mosquitoes were content to drink nectar for 3–5 min before backing out of the flower. When they did so, one of the sticky disks became attached to their eyes. As the mosquito backed out, she pulled out the stalk that connected the disc to one pollinium in the left or right anther chamber. She didn't like that sensation much and tried to wipe the disk away, but her forelegs really weren't built for such an operation.

When a mosquito yanks out a pollinium it stands up erect over the insect's eye, but not for long. As the stalk dries out it bends downward until it reaches an angle in which the mosquito may smear the wad of pollen onto a receptive stigma the next time it visits another orchid flower for a drink. Yes, Darwin found the same drying and bending mechanism in the butterfly-pollinated orchids he found near his country home. It's estimated that thousands of orchid species employ the same 'contrivance' regardless of which insect species pollinates the flower most often.

Len published his results in 1970, but did it really answer all pertinent questions? Were mosquitoes the only pollinators of this flower? How many mozzies fell for the trick a second time? How many were thirsty enough to repeat the behaviour on a second flower on a second plant and effect cross-pollination?

Obviously, Len and the bog weren't finished with each other. This time he worked with a faculty member, Dr Frederick Utech (1943–2001), and a piece of equipment known as a New Jersey Light Trap. The original device was invented in the late 1920s. Today, users hang it from a branch or mount it on a pole. Insects are attracted to its light at dusk and the battery-operated fan in newer models sucks them in depositing the little bodies in a collection jar at the bottom. This let both men harvest nocturnal visitors the following day without navigating such dangerous terrain at night. The new light trap worked even better than the dry ice method. While *Aedes communis* were most likely to wear the orchid's pollinia two additional species, *A. punctor* and *A. stricticus*, were also pollinarium vectors at a couple of sites. The big surprise, though, was the arrival of two species of small, speckled moths in the genus *Xanthorhoe*. They prefer colder latitudes in North America and their inchworm or looper caterpillars eat lady's bedstraw (*Galium*) or the leaves of local shrubs and trees. All 55 moths carried one to four pollinia on their eyes.

The orchid's mode of pollination was far from perfect. Despite the reward, the majority of mosquitoes and moths were not loyal and consistent visitors. The majority of insect visitors either failed to contact the sticky pads in the flower while drinking or were content with the nectar provided by one visit. While 38% of the flowers in Len's sites had at least one of their pollinia removed, only 14% of them set seed.

By 1971, Len left the blunt-leaved orchid behind and that left a most important question unanswered. How did mosquitoes find the orchid in bloom? It couldn't have been by colour as the greenish-yellowish petals were much the same hue as the surrounding vegetation. It's unlikely the plant exhaled carbon dioxide. Most healthy plants take in carbon dioxide with water and use it to trap photons of light to make sugar. That's the basis of photosynthesis. Diseased and dead plants release carbon dioxide as they rot and are broken down by microbes. Could there be something else in the living orchid's scent preferred by mozzies? People have smelled the flowers of the blunt-leaved orchid and described them as musky or reminiscent of a freshly mown lawn.

## WHAT DO DANGEROUS MOSQUITOES SMELL AND HOW DO THEY SMELL IT?

To answer this question, we had to wait another 50 years and the work of a team of eight scientists at the universities ofWashington in Seattle and California in San Diego. The blunt-leaved orchid also grows happily in the Pacific Northwest attended by more mosquitoes in the genus *Aedes*. You may wonder why the National Institute of Health was so happy to fund such a distinctly botanical project? The answer is that at least one species of *Aedes* is more than a greedy and annoying pest. *Aedes aegypti* is believed to have originated in Africa but has moved all over the world with the unwitting help of its preferred prey ... us. It's not happy in frigid regions, but that seems to be changing along with earth's climate. It was recorded recently in Canada. Often called the yellow fever mosquito, *A. aegypti* also transmits the viruses of Zika, Dengue, Mayaro and more. What could the orchid tell us about the mosquito's victimisation of humans? When the scientists introduced the yellow fever mosquitoes to the orchid's flowers for the first time, the noxious insects were more than happy to land and drink nectar.

Some simple experiments keeping the mosquitoes off the flowers with isolation bags showed how much the orchid needed its pollinators. In the absence of nectar-drinking mozzies, the pollinia usually went nowhere. Unbagged flowers exposed to visiting mosquitoes, or pollinated by hand, produced at least nine times the amount of fruits as in the bagged and isolated blossoms. The blunt-leaved orchid is obviously not very fertile in solitary confinement, It must have insect pollinators.

What was it about the scent that drew mosquitoes to the blunt-leaved flowers but not to the larger, far prettier flowers of the orange-fringed orchid (*Platanthera ciliaris*), bog candle (*P. dilatata*), slender bog orchid (*P. stricta*), Huron orchid (*P. huronensis*) and Yosemite bog orchid (*P. yosemitensis*)? If you are fortunate enough to find these wildflowers in bloom in a swamp or wet prairie in North America, you may see them visited by butterflies or bees

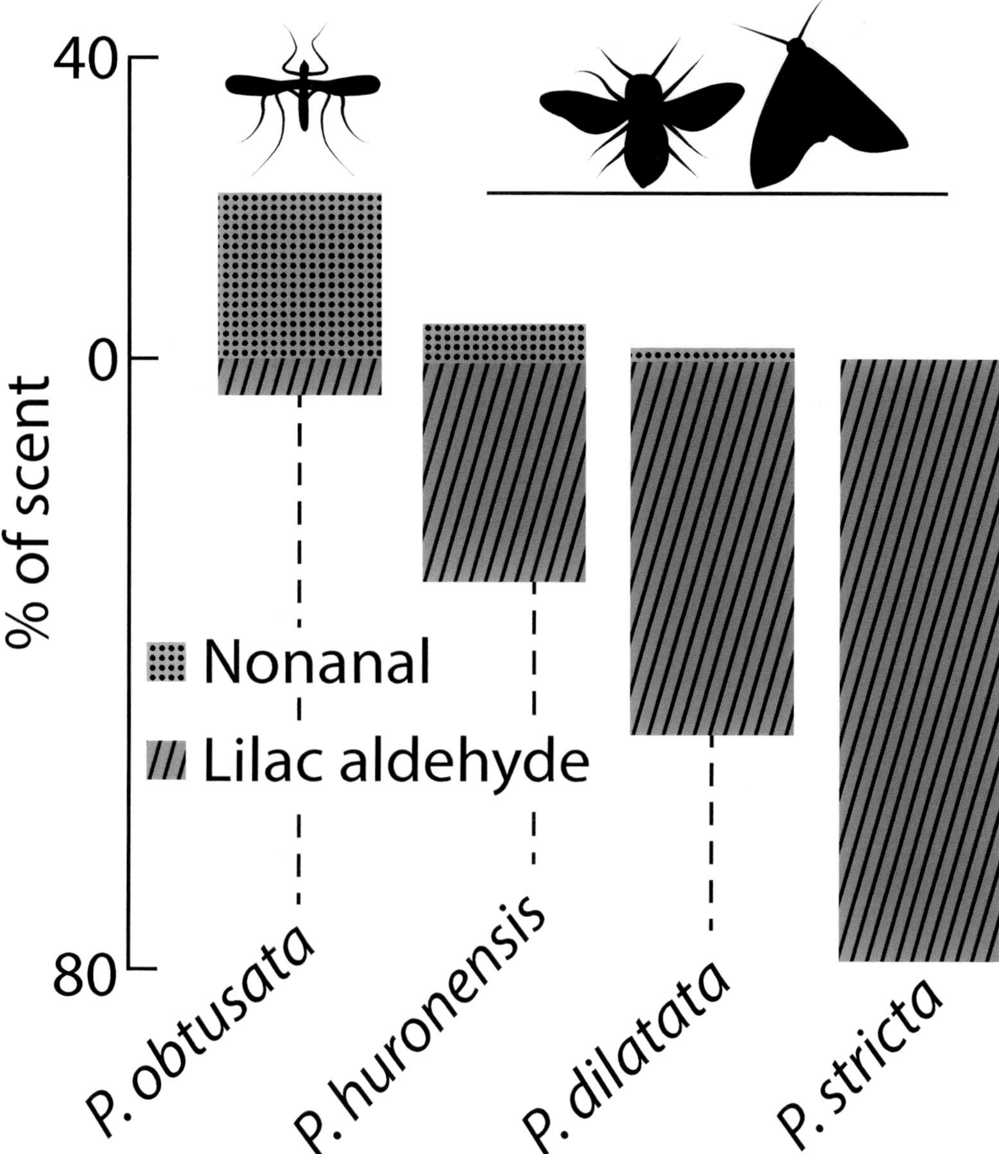


Although a total of 13 different molecules were detected in the flowers of the blunt-leaved orchid (*P. obtusata*) using gas chromatography and mass spectrometry (GCMS), 54% of the smell was based on nonanal combined with a dash of lilac aldehyde. While flowers of bee- or moth-pollinated *P. huronensis*, *P. dilatata* and *P. stricta* produced 7–8 identifiable compounds, 75–97% of their perfumes were dominated by lilac aldehyde with little room for nonanal. From Lahondere *et al.* (2019). (See Annotated bibliography.)

or even some of the larger, elegant, night-flying sphinx moths. Members of the scientific team subjected the scent of the flowers of all these relatives to gas chromatography and mass spectrometry. As usual, the components of scent in each flower were recorded as independent peaks along a graph. The flowers of each species made their distinctive odours by mixing half a dozen to a dozen different molecules, but in different concentrations. As we've learned, there's nothing unique about that. The big difference is that more than half of the scent produced by a blunt-leaved flower belonged to the molecule nonanal. The remaining five species secreted far less of this chemical or none at all. The scents of the flowers of the orchid species with larger and prettier flowers seemed spicier to human noses or at least reminiscent of preferred garden flowers like lilacs (*Syringa*).

Nonanal doesn't seem terribly special, at first. Lots of plants produce it. It's also found in the needles and scale leaves of several cone-bearing trees. It mixes with the fragrances of flowers made by other orchids as well as some jacarandas, carnations (*Dianthus*), Asian magnolias, members of the philodendron family (Araceae) and many others. The big difference is that it usually comprises only a fraction of the total body of chemicals that make up a flower's smell. It is such a minor molecule, in most cases, that the nose of a trained perfumer couldn't possibly identify or separate it in the blended fragrances of most tropical orchids.

## DOES THIS ORCHID SMELL LIKE US?

Mosquitoes, though, have no noses. They smell their world through different lobes that make up their antennae. The team of scientists found the appropriate lobe that connects with the mozzie's brain and makes it want to fly towards a strong source of nonanal. It's important to note here that it's the same lobe that makes the same insect fly away from DEET, the major component in most mosquito repellents. Why has natural selection promoted the rather obscure nonanal to signal the flower's presence to mosquitoes including the non-native and potentially lethal *A. aegypti*?

Nonanal is one member of a family of chemical scents known as aldehydes. It has a sister molecule known as 2-nonenal. This smell-alike molecule is also produced naturally and has

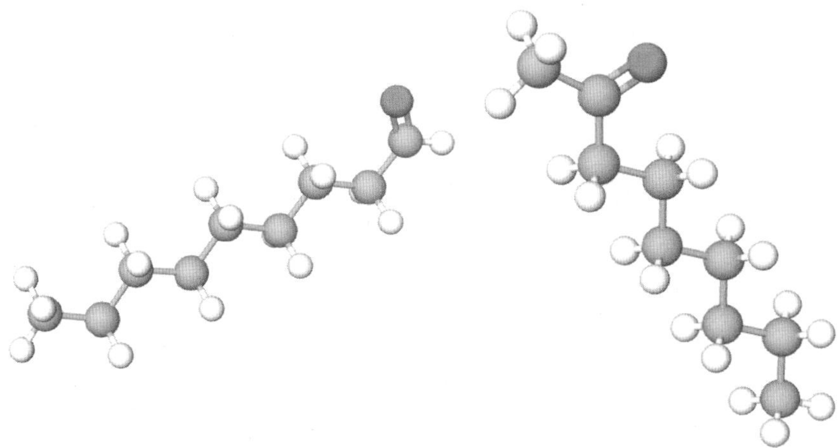

Two similar but unequal scent molecules. Nonanal (left) gives us nice floral or orange peel notes while we may describe 2-nonenal (right) as musty/greasy and it is also associated with the breakdown of fatty acids on our skin as we age. Hydrogen (white), carbon (light grey), oxygen (dark grey).

been found in beer and buckwheat products. People also describe 2-nonenal as grassy but there's something more we need to appreciate. While humans and insects may recognise the same scent, our human brains respond very differently to different concentrations. Some of us find 2-nonenal disgusting, complaining of a greasy odour.

Our skin gives off a blend of smells and not all of them are nice. In particular, some biochemists report that 2-nonenal is a dominant component in what we now refer to indelicately as old people's smell. Perhaps this explains the Middle Eastern tradition of older ladies washing themselves with the rose water they distilled at home. While the blunt-leaved orchid can't exhale carbon dioxide, its flowers can and do manufacture a fragrance similar to the favourite prey of certain mosquitoes. I'm not saying that 2-nonenal is the *same* thing as nonanal but, when blended naturally with a few more scents the mosquito obviously finds it alluring and maybe irresistible. Don't you enjoy aromas from the kitchen when someone you love is preparing your favourite, home-cooked meal? Some people involved in developing methods that better protect public health have suggested that mosquito traps be manufactured and baited with a combination of nonanal, lilac aldehyde and carbon dioxide. Widespread disctribution would be more economical, in the long run, then repeatedly dousing vulnerable populations, and homes, with DEET. Can't say I enjoy smelling DEET on myself when I must work in a field site close to a pond or swamp.

It also makes me want to speculate. The ancestors of today's blunt-leaved orchids must have evolved before our African ancestors migrated to far northern latitudes. At that time, mosquitoes of the Arctic Circle drank the blood of native mammals like bear, lynx, wolf, musk ox or reindeer. Furthermore, these orchids probably bloomed long before Eurasia's humans crossed into the Western Hemisphere via the Bering land bridge over 15,000 years ago. As humans moved in they much reduced the populations of both native predators and prey. Did this have a direct effect on the survival of the orchid? Is it possible natural selection encouraged a 'shift' in the mixture of scents made by the flower to make it smell more like us and less like the now extinct mammoth? If so, it's an imperfect solution. The orchids would produce more seeds if more mosquitoes visited more than one flower.

People living in the Northern Hemisphere today are far more numerous and far more likely to reach a ripe old age compared to our ancestors. Is evolution producing a contemporary mozzie that prefers pensioners to toddlers or teens? Once we retire, we have more time for those long summer walks. Could that influence the reproductive fitness of a little orchid with green flowers? Maybe this time we will find out in less than 50 years.

Of course, it's unlikely that the blunt-leaved orchid is the only plant pollinated by mosquitoes. We know that, in their search for sugar, male and female mozzies frequent many species with tiny, tubular flowers. I'm not suggesting that any mosquito at a flower is also a pollinator, especially when larger patrons like bees and butterflies are the most frequent visitors. However, people strolling along country paths, glades and meadows in North America in late summer are aware of how mosquitoes fly up as they brush against the flowering heads of snakeroots (*Agertina*), pye weeds (*Eutrochium*) and goldenrods all crammed with tiny floral tubes. It's also unlikely that the blunt-leaved is the only mosquito-pollinated orchid. It has other relatives in the genus *Platanthera* with small, green, flowers and they, in turn, are related to the Californian rein orchids (*Piperia*) with their stalks of tiny, green and white blossoms.

I deliberately wrote this chapter as more evidence that unique interactions among different organisms are not unique to Australasia. No one has found a mosquito-pollinated orchid in Australia, to date. I would, however, be remiss if I failed to introduce you to a clearly Aussie example of orchids pollinated by noxious insects. The evolutionary implications of this new relationship, in the next chapter, may be just as disquieting.

# 4

# Blood or tears?

*Fish gotta swim and bird gotta fly; insects, it seems, gotta do one horrible thing after another.*

– Pilgrim at Tinker Creek, *Annie Dillard*

WHEN AUSTRALIAN RAINS FALL FROM NOVEMBER TO FEBRUARY IN PARTS OF New South Wales and Victoria they stimulate the growth and flowering of a select number of summer wildflowers with drought-tolerant bulbs and tubers. This includes some summer orchids. There are at least 50 species of midge orchids in the genus *Corunastylis*. Most push up their flowering stalks following summer storms while others bloom on into mid-autumn in south-eastern Australia. A few species are quite rare, but others show the atypical habit of thriving where most native plants prefer to retreat. In New South Wales, while midge orchids are components of native shrublands and open she-oak (*Casuarina*) woodlands they also come up by their hundreds on degraded farmland, along eroded bush tracks and roadside verges. Further south in Victoria, they mix in with the weeds and European grasses in country parks and abandoned paddocks. They are native species that appear to thrive when local shires and small towns enact careful mowing policies.

While these small, native plants may flourish where people change the natural landscape, they don't receive much attention from most nature lovers.The Australian orchid expert David Jones called them 'living jewels'. He insisted there are at least 77 species, with many more to be described and named. Their flowers are measured in millimetres and you need a hand lens to identify the minute organs that comprise each flower, such as the blunt-leaved orchid in the previous chapter. Consequently, hobbyists and professional horticulturists of native plants aren't terribly enthusiastic about adopting midge orchids for container culture. I've attended plenty of orchid shows in Sydney and Melbourne over 40 years and admired pots of native rock orchids (*Dendrobium*), donkey ears (*Diuris*), and even the helmet and greenhood orchids we meet in Chapter 5, but can't remember a single potted midge orchid. While working in the Langwarrin Flora and Fauna Reserve, Victoria, I noted that dog walkers didn't give the three species of midge orchids one bit of attention.

I do agree with Jones that midge orchids have a certain *je ne sais quoi*, provided you look closely and wait for the next breeze. Unlike many members of the orchid family, midge orchids show a rarer and more conservative form of flower development. Most orchids make the stalk (pedicel) holding each flower bud twist 180 degrees before they open and expand. That means that the lip petal (labellum) becomes the lowest segment on the open flower and droops downward. When we look at the flower, the lip petal reminds us of an apron worn around your waist leaving your arms free to work. In most orchids, the lip petal is the pollinator's landing pad. The animal is guided into the flower by the colours and sculptures on the apron-like lip. It uses the lip to crawl **up** into the flower until its head or tongue is sandwiched in between the lip (below) and **under** the fused knob of sexual organs, known as the column (above). The flower holds the pollinator in place allowing the column's stigma to receive incoming pollen. Then the column releases a fresh pollinarium onto the pollinator as it backs out of the flower. It's the stereotypical mode of presentation most orchid flowers employ from the Wisconsin bog in the last chapter to the jungles of South America... or do they? Some Australian orchids do not twist their flower stalks.

The positions of the flower stalks of midge orchids and their Aussie relatives, the leek (*Prasophyllum*) and onion (*Microtis*) orchids, don't change at all. When the flower opens, the lip petal sits **above** the column resembling a cap, flag or curled plume. When the pollinator lands on the raised lip petal it must crawl **down** until its head or tongue lies in between its lip petal (above) and column (below). Once again, this permits the delivery of pollen to the stigma as the insect enters and the removal and dispersal of a fresh pollinarium onto the insect as it backs out. The difference is that many midge orchid species have added a mysterious novelty never seen in leek and onion orchids. The column ends in a chin and the lip petal connects to that chin by means of a thin band so flexible it is a hinge. Even a tiny lip petal creates some drag when it's propped upright at an angle. On a day with the mildest breezes the lip petals of the glandular midge (*Corunastylis filiformis*), fringed midge (*C. fimbriata*), elfin midge orchid (*C. archeri*), bearded midge (*C. morrisii*) orchids and others flap, flap, flap, up and down uncontrollably. One local naturalist, John Roslyn (Ros) Garnet (1906–1998), referred to this as the 'comically tremulous labellum'.

## ROS GARNET AND THE NATURALISTS OF VICTORIA

It's Ros Garnet we must turn to for the earliest and well told story of what was going on inside flowers of midge orchids. He published his findings in 1940 in the journal *The Victorian Naturalist*, back in the days when professional botanists regarded midge orchids as nothing more than tiny versions of the larger and more fragrant leek orchids. It was also a time in which the scientific names of plants reflected a more conservative approach to the diversity of the Australian flora. The scientific name of midge orchids changed twice in the 20th century. They've gone from *Prasophyllum* to *Genoplesium*, and not everyone is happy with their recent switch to *Corunastylis*. What is important is that the first half of the 20th century was a golden age for amateur naturalists in Australia. Hundreds of enthusiastic suburbanites and farmers made careful, long-term observations about plants and animals that were largely ignored by the few professional biologists employed by Australian universities and governments. People like Garnet reported their findings at monthly meetings of natural history clubs (most large towns had one) and were encouraged to publish their results in their society's magazine.

Garnet, while a great promoter of national parks, had no degrees in botany or entomology. He spent much of his professional life as a chemist in industrial and biomedical laboratories. Garnet also lived in an era before the majority of Australian plants were conserved. No one complained when he dug up wild orchids and transferred them to pots, which he cultivated for years. It was probably a wise move as he was stricken with polio at the age of 18, so a collection of potted specimens meant he didn't have to kneel too often or make daily hikes to field sites. Ros kept his collection outdoors and local pollinators entered and left at will. He learned that midge orchids appealed to some of the most deplorable residents of suburbia.

## FLIES VISIT GARNET'S POTTED ORCHIDS

We all know there are no summers in Australia without flies. While the bogs of Wisconsin and the Alaskan tundra has clouds of mosquitoes in summer, they represent only one unhappy portion of the order Diptera. By contrast, the Australian summer sees mosquitoes mixed with clouds representing many other fly species. As European colonists changed the environment to fit their concepts of landscape, sanitation and animal husbandry, a few opportunistic species benefitted. The most common remains the native bush fly, *Musca vetustissima*, in the blowfly family (Muscidae). They still invade picnics and barbecues, and are perfectly happy to suck up the sweat and secretions on exposed skin. Bush flies benefitted from management of herds and flocks as the dung of hoofed animals established maggot nurseries, far more generous than the droppings of native marsupials. Then we have the much loathed, biting midges that locals refer to as sand flies. These bloodsuckers haunt some of the most beautiful coastlines and most belong to the genera *Culicoides* and *Leptoconops* in the family Ceratopogonidae. After a pleasant day at the beach you may come home with little, itchy red welts made by insects only a few millimetres in length.

Neither of these pests visited Garnet's pots, but there had to be a pollinator as he noticed that a few of his flowers set fruit. One March day in 1934, he admired a spray of bearded midge orchids bearing 14 open flowers when he noticed tiny flies visiting them. These insects, he noted, were on the lip petals browsing 'in a leisurely fashion'. They were so oblivious to everything, except the flowers, that he could lift up the flowerpot and follow the flies' movements with a hand lens. Each fly crawled down the upright lip. Based on one of his hand drawn illustrations, it looked like that flapping petal bent down slightly in the direction of the column under the sheer weight of the insect. The fly found a drop of nectar to drink at the base of the furry and decorated lip petal. Garnet, ever the chemist, lowered a large test tube over a whole flowering stem infested with little flies. When three of these flies backed out of three flowers, each insect wore a pollinarium on the back of its thorax. As in almost all orchids in which the flower has only one pollen-making anther, the entire pair of pollen balls (pollinia), attached to the same stalk (stipe) and its sticky plug, became attached to the fly's back only as it backed out of the upside-down column.

According to Ros, these flies had preferences according to the orchid species growing in each pot and the bearded midge was the clear favourite. The elfin and dark midge orchids (*Corunastylis nigricans*) took second place. He observed that the flies ventured into the flowers of the sharp midge orchid (*C. despectens*) only after they'd had their fill of bearded and elfin midges. Considering the size of the pollinia, the little pollinators carried quite a burden. They were only 2 mm in length and, as a dedicated fact checker, Garnet used the most sensitive

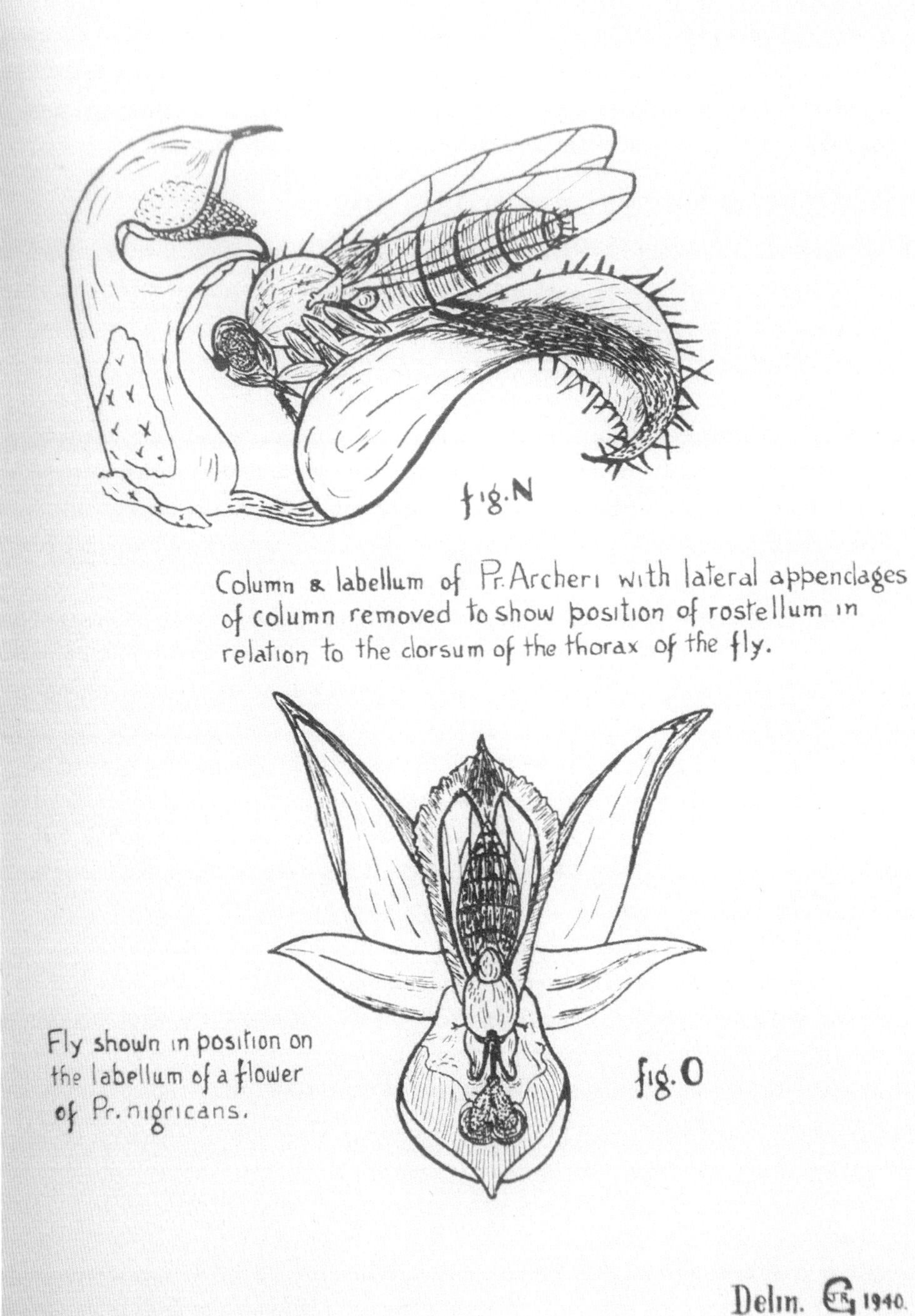


Ros Garnet's 1940 drawing of a chloropid gnat taking nectar at the base of the lip petal while the back of its thorax faces the orchid's column. There is one thing wrong with this picture. Garnet should have inverted the flower and insect (pointing lip petal and fly abdomen upwards) instead of presenting them horizontally. Reproduced with the permission of the Field Naturalists Club of Victoria.

balance he could find and weighed his flies. He calculated that the fattest fly weighed only 0.05 milligram.

Who were these little pollinators with such sluggish feeding habits? In March 1937, Garnet found a comparatively larger and chunkier fly in the flower of a bearded midge, growing wild in the bush. It seemed stupefied and had to be withdrawn with a pair of forceps, but it was not carrying pollen wads. An entomologist at the Melbourne Museum identified it as one of many fruit or vinegar flies (*Drosophila*). Garnet also sent specimens of his previous catches, the ones wearing pollen wads, to AL Tonnoir, an entomologist working at the Australian Council for Scientific and Industrial Research in Canberra. These flies were very different and belonged to the genera *Caviceps* and *Oscinisoma* in the family Chloropidae.

In Australia, chloropids are best known as eye gnats or eye flies. They prefer to drink tears from animal eyes, but they're not picky. It's common for some species to land on an animal's anus first and then proceed up to the host's face. They are also fond of weeping sores and fresh lesions. Fortunately, chloropids are not identified as carriers of human diseases in Australia yet, but there are species in the tropical Americas known to carry unpleasant bacteria and viruses. It should be of no surprise they transmit the microbes of acute pinkeye (conjunctivitis), various skin infections (including yaws) and even Brazilian purpuric fever. Like the *Aedes* mosquito in Chapter 3, Garnet's orchid flies came from a bad family, but were happy to drink plant-made treats when offered.

While hover flies, tangle vein flies (Nemestrinidae) and the various families of fungus gnats we will soon meet contain thousands of flower-visiting species all over the world, references to eye flies as pollinators are few. In the Western Hemisphere they are among the flies trapped temporarily in the grotesque floral sacs of birthworts (*Aristolochia*), which are unrelated to orchids. Chloropids are also attracted to scents secreted by some lantern or parachute flowers (*Ceropegia*) in the dogbane family (Apocynaceae). They may also play some role in pollinating flowers of the American juniper mistletoe *(Phoradendron juniperinum*), but that's about it. This time, Australia offers the most convincing case for pollination by an atypical pollinator.

The interest in eye flies and the sex life of midge orchids didn't stop with Garnet, although there have been long gaps. In Australia, the Tuncurry midge orchid *Corunastylis littoralis* (previously *Genoplesium littorale*) is now restricted to a few dunes by the sea, as they are victims of sand mining. A team of three scientists, led by Dr Colin Bower, looked at reproduction in this wildflower. Despite the reduced populations the survivors enjoyed a long flowering season (February–May) and a healthy rate of fertility with slightly less than half of the flowers on stalks (~44%) maturing as seed-filled fruits. The orchid was visited by five species of chloropid flies. Most of them were females belonging to the genera *Conioscinella* and *Cadrema*. They were doing their job as almost three-quarters of all the flowers on stalks examined had their wads of pollen removed. While these eye flies were the true pollinators, the flowers were also probed by jackal flies (Milichiidae) and they are thieves. Their primary food of jackal flies is the blood or haemolymph of insects and other invertebrates wounded or killed by spiders and predatory insects, such as robber flies, mantises and assassin bugs. Ecologists call this kleptoparasitism. It's the same process on a much larger scale when lions take prey killed by wild dogs or hyenas, and frigate birds or eagles relieve seabirds of the fish they've caught.

As eye flies and jackal flies can't turn down wounds, sores or lesions, could the nectar in the Tuncurry orchid mimic insect body fluids? It's hard to say. The flowers should smell like dead or dying insects but most midge orchids are so small people don't notice their scent. A few have recognisable fragrances reminiscent of sour milk or lemon peel, not the stink of bedbugs or a squashed cockroach. Scent must play some sort of complicated role as wildlife photographer Rudie Kuiter recorded other insects on some midge orchids carrying the pollen wads. This includes some minute members of the family Scatopsidae, known commonly as dung flies, so you know what they prefer to smell.

## EYE FLIES AND MIDGE ORCHIDS, OUR EVIDENCE

Now it was the turn of a team of four people representing Australia, America (me) and China to conduct a series of field studies over two years. The Australian Orchid Foundation funded our field study in New South Wales and we went to work in 2016 collecting pollinators and bagging and tagging flowering stems. We later added Dr. Qi Qiao, a visiting Chinese botanist, to analyse pollinated pistils I brought back to my lab in St. Louis. Field team member Dr Zong-Xin Ren came to our field sites from the Kunming Institute of Botany in Yunnan, China, carrying some hand-held devices that could suck up and trap the perfumes of the tiniest flowers. Team member Wendy Grimm, then a PhD student at Macquarie University, NSW, was well acquainted with midge orchids as they shared the same sites, and some of the same habits, as the rare orchid she was studying.

No one had sharper bush sense and skills than Wendy, or paid a greater attention to detail. She'd made a singular discovery before we arrived in Sydney. The exaggerated flowers of the fringed midge orchid (*Corunastylis fimbriata*) were making droplets. However, they didn't appear at the base of the lip petal, unlike the species studied by Garnet or Bower and his team. Each flower of fringed midge made two droplets. Each droplet was secreted at the tip of each of the two wing-like structures attached to either side of the column. We soon saw that chloropid flies were as happy to drink wing lobe fluids as their relatives had been to suck from the lip petals of flowers studied by previous authors. The adaptation was not unique to the fringed midge either. Our second Australian partner, Brian Towle, first observed it in the glandular midge orchid (*C. filiformis*). I don't remember who detected it next in the flowers of Rupp's midge orchid (*C. ruppii*).

When Charles Darwin studied orchid flowers in the 19th century he came to the conclusion that their evolution followed a path of organ reduction. The series of developmental studies that followed, long after his death, have largely proven him right. Darwin noted that if an orchid flower made only one fertile stamen, and most of them do, then the same column or rod of sexual organs usually bore a pair of tiny wings or ear-like appendages that fused to either side of the single good stamen. Ultimately, the two wings and the good stamen fused together and then all three fused to the neck of the pistil. Darwin came to the conclusion that these two wings were reduced sterile stamens and most botanists still call them staminodia comparable to the larger, solitary staminodium found in the lady's slipper flowers in Chapter 2. Darwin believed that staminodia were largely vestigial organs, like the human appendix, and no longer

Column wings of the flowers of the fringed midge orchids (*Corunastylis fimbriata*) secrete large droplets. Modified with permission from Ren Z-X, Frim W, Towie B, Qi Q, Bernhardt P (2020) Comparative floral traits in Corunastylis (Diurideae; Orchidaceae) with novel applications: do some species bleed or blink? *Muelleria: An Australian Journal of Botany* **39**, 27–38. doi:10.5962/p.340568

performed any useful functions. In contrast, pollination biologists are learning gradually that the staminodia of some orchids may be adaptive. Australian sun orchids (*Thelymitra*), for example, have unusually complicated, colourful and sculptured staminodia that attract bees that cling temporarily to these outlandish knobs and brushes. The two wings of the column of each greenhood flower in the next chapter form two walls that help hold the fungus gnat prisoner after the lip petal springs into action. In the male flowers of tropical American *Catasetum* orchids the staminodia are long, sensitive antennae. If a bee touches either of them, the triggered flower shoots its pollen wads onto the insect's back.

After our fieldwork ended the first year, I sent flowers pickled in alcohol back to my laboratory for dissection and a much closer look. The shapes of the column wings of six species of midge orchids varied, but they all shared two things in common. First, each wing divided into two lobes. One lobe was rather hairy, but didn't secrete fluids. This fuzzy thing was the staminodium proper. The second lobe had smooth margins and this is a structure most orchidologists call the auricle, as it makes the fertile stamen look like it's wearing a pair of fox or bat ears.

Second, under a microscope, this paper-thin auricle contained one to more than 40 additional cells filled with spikey, blackish crystals known as raphides. Most of the cells containing raphides were densely packed into the tip of the auricle. When orchids pack raphides into their cells it usually means the tissue or organ will secrete some sort of slime. Were our eye flies drinking a weak slime or a true sugary nectar? We couldn't tell. While the auricles were obviously secreting enough droplets to keep the pollinators happy, there was very little fluid to gather and it was rather viscous. I could never collect enough in a capillary tube to deliver to my portable sugar refractometer. In the beverage industry, these simple devices are used to record the percentage of cane sugar (sucrose) dissolved in soft drinks.

The benefit of staminodia secreting droplets in our three species of midge orchids was obvious. Eye flies gorging on this reward were drawn even closer to the parts of the column that accepted and released pollen wads. Sometimes we found flies so glutted that they continued to perch on parts of the flower while they regurgitated fluid bubbles, as flies do, and this was to the benefit of the plant. As Darwin also noted, when an insect backs out of an orchid flower wearing pollen wads the stalk or stalks that separate the pollen wads from their sticky connective plug (viscidium) are still turgid and the pollen wads are held almost erect to the insect's body. In this fresh position at slightly less than 90 degrees the wads won't contact the receptive but angled stigma of the column in a second flower. That's useful if the pollinator visits more blossoms on the same stalk as it prevents self-pollination. But how does cross-pollination occur then when a fly picks up pollen wads on one plant and then flies to a second?

The answer is that those wads don't stay in the semi-erect position forever. Once exposed to fresh air pollinarium stalks start to dry out and bend down within a few minutes. We saw

Three female eye flies (*Conioscinella*) carrying the pollinaria of midge orchids. Note how the stalk (stipe) of each pollinaria bends down until the yellow pollinia are just above the fly's head. Photo by John Martin.

the same process occur on the mosquitoes carrying the pollinaria of the blunt-leaved orchid in the previous chapter. This repositions the pollen wads at an angle that converges with the horizontal stigma surface of a second flower. The progress of drying stipes was dramatic in our midge orchids. They bent forward and so low that the pollen wads often rested on the bristles on the fly's head. Under natural conditions the pollen wad is now in a position to rub against the sticky surface of the column's stigma while the fly drinks droplets from the auricles. Midge orchid pollen wads are crumbly and leave packets of sperm-filled grains on the stigma.

We had to collect lots of visitors to the flower to determine which ones were most likely to carry pollen wads. Eventually, we amassed 323 insects over two seasons, from flowers of four midge orchid species. We also collected eye gnats that were lounging on twigs and grass blades adjacent to our orchid flowers. As Garnet first noted, eye flies are in no rush and spend hours on the same plant. Nearby vegetation was a perfectly good place to digest auricle secretions.

Most of the true flower visitors we learned were female eye flies. The males were most likely to hang around on orchid stalks, twigs or grass blades. The males rarely entered the orchid flowers. Only two males out of the hundreds of little bodies we examined under the microscope carried pollen wads. It looked like the droplets were providing something the females needed to complete their lifecycles. As in Garnet's study, our female eye flies had specific tastes, as it was common for more than one midge orchid species to bloom along the same trail or patch of ground. Given a choice, the females didn't much care for flowers of Rupp's midge orchid. I noted that the auricles of this flower didn't make as much fluid and their lip petals didn't flap enticingly in the breeze.

How much room is there for pollen wads on the back of a fly that's less than 3 mm in length? No fly caught carried more than two pollen wads from two flowers on her thorax. I still wonder how she was able to launch herself into the air while wearing two balls and chains? I came to the false conclusion that an eye fly wearing two pairs of pollen wads apiece could no longer fly at all and was trapped on orchid stalks until she perished. I was wrong. Wildlife photographer Rudie Kuiter observed a single eye fly wearing three pollinaria at one site in Victoria. She opened her wings and buzzed off without obvious difficulties.

The big question now was, how many eye fly species did we collect and did each insect species prefer a different species of midge orchid? We handed all our specimens to two entomologists at the Australian Museum in Sydney. The good news was that all the flies carrying pollen wads belonged, once again, to the genus, *Conioscinella*. The bad news was the warning we received from the entomologists. No one had worked on the classification of *Conioscinella* species much since the 1920s. The best they could do was to compare our specimens and subdivide them into morphotypes.

This is a common procedure in Entomology when an insect lineage hasn't been studied for decades and museum cabinets are packed with unnamed specimens. It may take years to describe and publish a ponderous monograph revising the original classification and describing new species. It was far easier and quicker to check all of our 323 corpses and subdivide them into a discrete number of groups based on shared characters like body dimensions, number and position of bristles, vein maps in their wings, leg colours, and so forth. They were categorised as morphotypes and put into storage. Someday, we hope a new student of insect diversity will use our collections, and others amassed in museums all over Australia, to write a welcome treatise on the species of eye flies of the antipodes.

We were informed we'd collected six, different morphotypes of *Conioscinella*, but there was a catch. There were only 10 specimens representing five of the morphotypes. That's right, ~96% of all the flies we trapped belonged to the same morphotype and, most probably, the same species. That means that over two summers, four species of midge orchids with overlapping flowering periods, often blooming at the same times and sharing the same locations, may be in competition for the exact same pollinator. This may be a common problem for midge orchids in some parts of Australia.

You may ask, how did we collect such tiny insects? Butterfly nets won't do for bodies a couple of millimetres in length. The New Jersey Light trap is of no use when your insects are active during the day. They get lost in the folds of the net and escape through rips in the weave. We turned to a small, but effective invention available since the 1930s by the American entomologist Frederick William Poos Jr. Technology has improved the materials that make up the insect aspirator, or pooter, but they all work on the principle that nature hates a vacuum. It's a small, capped vial with two holes for two tubes on the cap. First, you point the shorter, harder, horizontally held tube at a tiny insect. Then you suck in air through the second, longer rubber hose. Your prey should end up in the vial. We killed our daily collections in a home freezer. Was there a danger of inhaling the insect? We had some of the better models in which the hole under the cap for the rubber house was blocked by a fine mesh screen.

I happily roamed the bush 'pooting' up dozens of its citizens at a time, until the mesh in my vial fell out. It was not at the best moment. One of our sites featured additional summer wildflowers. I was intrigued by the insects visiting the dainty, white-and-green flowering heads of flannel flowers (*Actinotus*). Some eye flies visited the flannel flowers, but so did some of the tiniest native bees. Without the mesh I sucked a bee into my mouth and was fortunate not to be stung. Pooting also showed us that eye flies weren't alone on the midge orchids. I returned from our sites to find little spiders in the vial. One had the time to make a little web and was eating one of the flies.

A pooter gives you a different perspective while you hunt. You don't disturb the vegetation as you would if you whack away at it with a net. We learned that our orchid flowers were the centrepiece in several circles of life. The eye flies that fed on them were prey to several spider species including a beautiful green lynx (*Oxyopes*). As Australia is a country of many flies it shouldn't surprise us that it's a place where many plants eat flies. Below the stalks of fringed midge at our Bloodwood Road site in Arcadia, NSW, were rosettes of sticky leaves. They belonged to two species of sundew (*Drosera*) and I found the remains of *Conioscinella* on their leaf glands. Then there was the morning a big, fat, flightless grasshopper began eating every fringed midge flower on the same stalk. She wasn't alone either, and was allowing a smaller male to mate with her. I pickled the two of them *en copula,* sent them to a specialist in Canberra who identified them as *Exarna includens.*

## COLLECTING AND ANALYSING MIDGE ORCHID SCENTS

Ren harvested fragrances from three midge orchid species with an oven bag (used more often to roast a turkey), an absorbent trap packed with glass wool, and a battery-operated PAS-500 vacuum pump with Tygon tubes. Scent analyses, while intriguing, didn't give us the complete story we wanted. Like the eye flies we observed and gathered, the people in the plant chemistry laboratory at the Kunming Institute of Botany were in no rush. By the time our scent samples came out of long-term storage I'm afraid several volatile molecules had broken down or dissipated to the point they couldn't be identified. In bloom, the fringed midge smells strongly and pleasantly like lemongrass used to flavour dishes in Vietnamese restaurants. The Kunming chemists should have detected citronellal or citronellol, but they were not detected in our samples.

Six molecules were detected and the relatively unloved Rupp's midge tested positively for five. This included tridecane, which was exciting as the same perfume attracts *Conioscinella* midges to lantern flowers mentioned earlier. It is also a defensive chemical used by some stinkbugs. This could add to the earlier theory that some midge orchid flowers are pretending to be wounded or dying insects to attract ravenous flies, but that's not a smoking gun. Tridecane has also been detected in the perfumes of other orchids, and unrelated wildflowers representing 35 more plant species that are definitely not pollinated by chloropids. Likewise, Rupp's midge and the glandular midge also secreted 8-Heptadecene and that's an intriguing clue. The flowers of the commercial cocoa bush mix 8-Heptadecene and tridecane with other scent notes to attract their major pollinators – pregnant biting midges in the family Ceratopogonidae.

The big surprise was nepetalactone secreted exclusively by the fringed midge. This is usually harvested from stems and leaves of catnip (*Nepeta cataria*) in the mint family (Lamiaceae). The molecule lures cats, but in commercial doses it also repels those awful *Aedes* mosquitoes discussed in Chapter 3. If dozens of midge orchid species compete for the same pollinator, over the same flowering period, one wonders if the winning flowers playing the pollination game are the species that make the scents that most appeal to the most common eye flies?

## HOW FREQUENTLY DO MIDGE ORCHIDS SET SEED?

One thing was obvious. When deprived of eye flies, reproduction in three midge orchid species flopped. We isolated the flowering stems of three species under organza bags while still in bud. When the buds opened the flies couldn't reach them through the mesh. Some flowers tried to self-pollinate and we found a few pollen grains on the sticky surfaces of the stigmas and a few sprouted tubes into the pistil necks. It didn't really matter in the long run as no bagged flower ever set fruit.

It was a different matter when eye flies had complete access to flowers of their choice, but fruit set rates varied according to year, location and midge orchid species. At Crib Point, Victoria, over 90% of the flowers on stalks of the eyelash midge orchid (*Corunastylis ciliata*) had swollen ovaries. Rates of ovary fertilisation were more modest in the populations of the three species we monitored in New South Wales, ranging from 38–43%. This, in fact, is more typical of cross-pollinated orchids that produce lots of small flowers on their stems while offering some sort of reward to their pollinators. While Rupp's midge orchid was a less popular flower for eye fly visits, it appeared to be just as successful converting flowers into ripe fruits as the glandular and fringed midges, but Rupp's midge suffered from a lack of quality control. When Ren and Wendy compared the development of embryos in seeds in each fruit. Only 42% of the tiny seeds in Rupp's midge had healthy, fully formed embryos compared to 70% in the glandular midge and more than 50% in the fringed midges. Perhaps flies on Rupp's midge lingered too long, pollinated more than one flower on the same stalk, with pollen from the same plant and, as usual, inbreeding led to more seeds containing genetic errors.

What we can say is that the system, as first described by Ros Garnet, works well. Although not every species of midge orchid flower pleases the taste of the same eye fly in the same way every year. What we can't say with any precision, at this time, is how this system works. Are all the midge orchid species trying to appeal to insects with the most ghoulish tastes? Do these flowers really secrete aromas reminiscent of fatal injuries or slow exsanguination? Is the flapping lip petal produced by many, but not all, midge orchid species some sort of decoy mimicking the throes of a dying insect? Do the flowers release an ersatz aroma of dripping gore? Yes, the lip petal leans a bit as the fly crawls down to drink its reward, but I never saw it dump the pollinator onto the waiting column like the dunking chair at a carnival. There's no automatic spring action here like the lip petal in the greenhoods in the following chapter.

## OUR WEEPING FLOWER HYPOTHESIS

My multinational team of botanists and entomologists thinks there may be a second explanation that dovetails with the first.one proposed by Bower and his people. The surface of the lip petal is dark and, depending on the species, its margins are often lined with hairs or bristles. The longer wing lobes that are attached to the dripping auricles are also fuzzy and furry. We know what an eye fly wants to drink. Maybe the flower is a small-scale mimic of a mammal's eyelash sending a false message by pretending to blink for its prospective pollinator as the lip petal flaps up and down. If we are ever able to analyse the nectar we can check for amino acids and organic salts expected in the eye secretions of marsupials or any other vertebrate. Who says a flower can't cry vegan tears?

We have looked at tiny flies pollinating orchids under the glare of the Australian summer sun. There's another side to this story during the Australian winter found in the next chapter. It's full of deceit, so it's not surprising much of it occurs in the shade.

While tiny fossil flowers of water lilies are at least 90 million years old (Cretaceous) each bud of a massive and modern giant water lily (*Victoria*) from South America opens, blooms and dies within less than 24 h. Photo by Retha Edens-Meier.

This amber fossil of the bee-like crabronid wasp *Discoscapa apicula* was found in a mine in Myanmar and is estimated to be 100 million years old (Cretaceous). Photo by George Poinar.

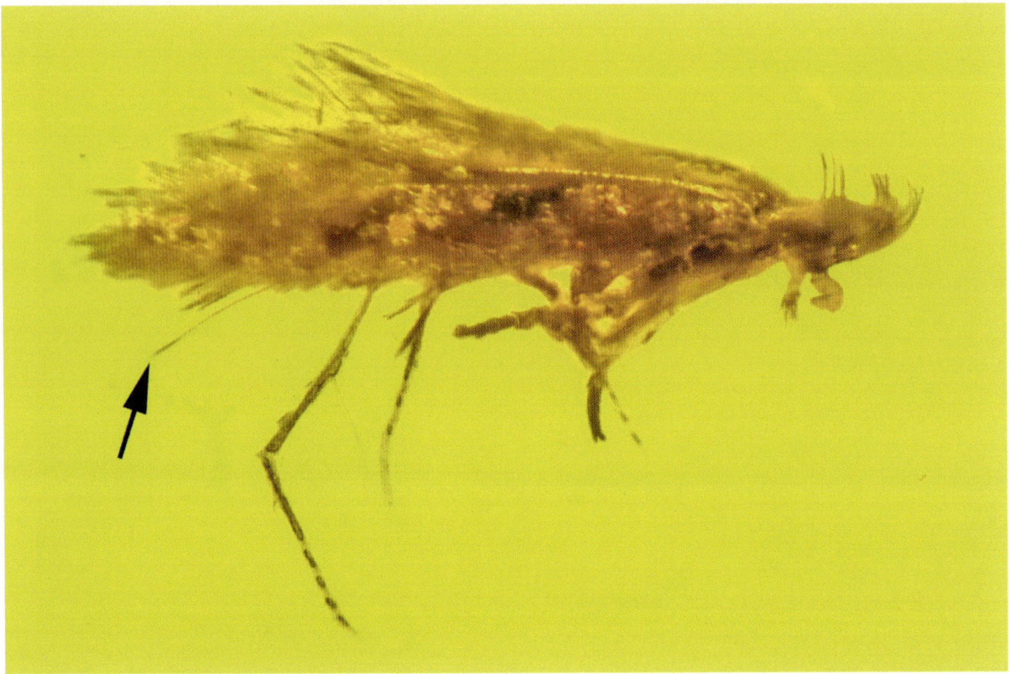

Amber from Myanmar, estimated to be 97–110 million years old, made this fossil of *Tanyglossus orectometopus,* The moth had a thin tongue almost as long as its body (black arrow), so it probably sipped nectar from ancient flowers. Photo by George Poinar.

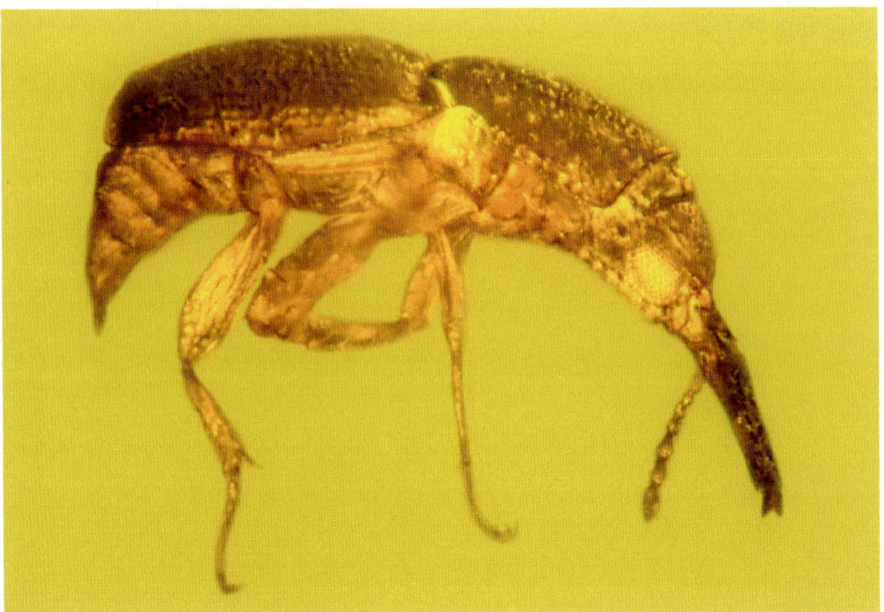

Amber fossil from Myanmar of *Furcalabratum burmanicum* (length 2.6 mm), a short-winged flower beetle (Kateretidae) from the mid-Cretaceous (97–110 million years old). It is in the same family as *Pelretes vivificus.* Photo by George Poinar.

The 'motherly' flower of queen or showy lady's slipper (*Cypripedium reginae*). It has the second largest lady's slipper flowers in Canada and America and is pollinated by medium-sized bees. Photo by Retha Edens-Meier.

This carpenter bee (*Xylocopa virginica*) is too big to fit inside the slipper of the 'maidenly' yellow lady's slipper (*Cypripedium parviflorum*). Photo by Retha Edens-Meier.

This endangered *Cypripedium lichiangense* of Yunnan is one of the largest crone flowers. Its long-lived, fuzzy, brownish 'toilet bowl' mimics rotting leaves and humus attracting pregnant hover flies (*Ferdinandea cuprea*). Photo by Zong-Xin Ren.

The degree of inflation of the lip petal of a lady's slipper orchid varies within its own species and between different species. Compare the lip petals of the motherly *Cypripedium fasciolatum* to a chicken egg (top row) and then compare the top row lips to lips of natural hybrids between *C. fasciolatum* and *C. tibeticum*. Photo by Zong-Xin Ren.

A cluster of flowers on a stalk of fringed midge orchid (*Corunastylis fimbriata*) featuring a female chloropid gnat (*Conioscinella* species) wearing the orchid's pollinia. Photo by Zong-Xin Ren.

A flowering stalk of a fringed midge orchid (*Corunastylis fimbriata*) smells like lemongrass. The fringed lip petal (located at the top of each flower) flaps during the slightest breeze. Photo by Zong-Xin Ren.

A female fungus gnat (*Exechia* species) wears the pollen wad of the Chinese helmet orchid (*Corybas shanlinshiniensis*) on the back of her thorax. Photo by Zhou-dong Han.

The flower of the Chinese helmet orchid (*Corybas shanlinshiniensis*) waits for fungus gnats. Photo by Zong-Xin Ren.

Two dagger flies bring dead insects (nuptial gifts) to compete for the deceptive lip petal of a bearded or plumed greenhood. Photo by Rudie Kuiter.

Two male fungus gnats are attracted to the cocked and ready lip petal of a greenhood (*Pterostylis*). Photo by Rudie Kuiter.

With the greenhood orchid's trap sprung, the male fungus gnat is now immobilised in between the wings of the column's style. Note how the white glue gun of the column is pressing against the gnat's thorax but it is also connected to the yellow pollinia. Photo by Rudie Kuiter.

A male fungus gnat is poised directly on the dummy female of the cocked and ready lip petal of a rusty greenhood (*Pterostylis* species). Photo by Rudie Kuiter.

The first day the homely flower of Brown's peony opens it is in the female phase. The green tips of the carpels are ready to receive pollen but the anthers in the same flower are shut promoting cross-pollination. You can barely see the green nectar glands between the anthers and red ovaries of the carpels. Photo by Andrew Huber.

A wasp queen sticks her head and thorax into the nodding flower of Brown's peony to drink nectar. Photo by Andrew Huber.

In some Australian spider orchids, the dummy female pattern on the lip petal is augmented by a pair of fake eyes towards the base of the column neck. This possibly helps to assure that the male wasp will always face the fake eyes positioning his thorax directly under the 'glue gun' under the tip of the column. Photo by Rudie Kuiter.

The elaborate collection of blackened warts and skin spots and hairs on the lip petal of the hammer orchid (*Drakaea livida*) lures male wasps. Photo by Mark Brundrett.

A male wasp inspects the lip petal of a carousel spider orchid (*Caladenia arenicola*) in Western Australia as he is still copulating with a wingless female of his species! Photo by Mark Brundrett.

Look at the number of pollinia on the back of this fungus gnat in the family Sciaridae. It has obviously visited more than one flower of the mayfly orchids (*Acianthus caudatus*) in a winter woodland, encouraging cross-pollination. Photo by Rudie Kuiter.

A male thynnid wasp wears the pollinia of a spider orchid on the back of his thorax. Photo by Rudie Kuiter.

Note how the head of the male thynnid wasp contacts the column of the spider orchid (*Caladenia*) as the wasp's claws grab the lip petal. Photo by Rudie Kuiter.

A male *Campsomeris* wasp attempts to copulate with a bearded orchid (*Calochilus*). Photo by Rudie Kuiter.

This is how a mating flight should end for thynnid wasps. The male perches on a flower that makes nectar, like this leek orchid (*Prasophyllum*), allowing the wingless female a drink before he releases her to return to the soil where she will lay her eggs in curl grubs. Photo by Rudie Kuiter.

A South African monkey beetle grazes in the floral cup of a *Gazania* daisy. Note the black and white 'beetle marks' on the ray petals. Photo by Steve Johnson.

A heliozelid moth perches on a broad, four-lobed stigma of a brown boronia flower (*Boronia megastigma*). Globs of moth-deposited pollen on the stigma are visible in the flower in the lower right-hand corner. Note the prominent black staminodes alternating with the almost hidden fertile anthers releasing their pollen beneath the stigma lobes. Photo by Liz Milla.

The white flowers of Spanish bayonets and Joshua trees (*Yucca*) are deceptively simple with its sepals and petals are converted to tepals and six stamens surround three carpels fusing into a pistil. Nectar glands are absent, so would you expect them to be visited by moths? Photo by Joey Santore – see iNaturalist.

A captive tube-lipped nectar bat (*Anoura fistulata*) of Ecuador shows the length of its tongue. Photo by Nathan Muchhala.

*Anoura fistulata* at the wagon wheel flowers of an unidentified *Marcgravia* species. Notice the long echolocation bracts dangling from the flowering pinwheel. Photo by Nathan Muchhala.

*Anoura geoffroyi* hovers as it takes nectar from a flower of *Meriania tomentosa*. Photo by Nathan Muchhala.

# 5

# Gnats in a winter bouquet

*Where are the songs of Spring? Ay, where are they?*
*Think not of them, thou hast thy music too,—*
*While barréd clouds bloom the soft-dying day,*
*And touch the stubble-plains with rosy hue;*
*Then in a wailful choir the small gnats mourn*
*Among the river sallows, borne aloft*
*Or sinking as the light wind lives or dies*

– Ode to Autumn, *John Keats*

At the age of 19, Dorothea Mackellar (1885–1968) felt she'd been visiting England and Europe too long and was feeling homesick. She wrote a poem, *My Country*. Generations of Australian children are still expected to read it because it extols the natural wonders and challenges of their continent. It belongs to an era when authors and artists encouraged Australians to develop some degree of national pride, or at least a sense of proportion. One stanza has always attracted my attention, in particular and goes, *Green tangle of the brushes, / Where lithe lianas coil, / And orchids deck the tree-tops / And ferns the warm dark soil.* Oh Dorothea, you were such a city girl (born and raised in Sydney). Why look up into the branches? Most of Australia's 1,300 orchid species grow down in hard, shallow soils like the midge orchids in the preceding chapter. Many bloom in winter.

That was a lesson I learned fast during my first field trip to the countryside around Melbourne back in August 1977. My new advisor, Malcolm Calder, promised a Saturday drive during the last month of the Aussie winter. The sky was overcast and the *Eucalyptus* woodlands around Moe were cool and damp. Malcolm pointed out the first clump of nodding greenhoods or cow-horns orchids (*Pterostylis nutans*). I'd never seen them outside of photographs in a book I'd purchased while a university student in New York. It made me feel like I was a tourist in a Los Angeles delicatessen seeing my first Hollywood star finishing her pastrami sandwich, and I was hooked. While my PhD studies focused exclusively on the sex lives of totally unrelated plants, I was resolved to find and identify every orchid species native Australia. That turned into a lifetime of commitment.

During those early years I spent looking down on the woodland floor and among low, open shrubs, known as heaths. I noticed that the forms and colours of native orchids segregated as the seasons changed. The flowers made by the orchids of mid-late spring came in a variety of whites, yellows, purples and blues, and that included the spider caladenias we will encounter in Chapter 7. The orchids of winter to early spring were more likely to tint themselves in restrained shades of green, rusty-brown, dull red or iodine, and that included my nodding greenhoods. Conservative and formal colours remain the rules for winter greenhoods, and mayfly (*Acianthus*), gnat (*Cyrtostylis*) and helmet or cradle orchids (*Corybas*). How do we explain the seasonal drabness in flowers belonging to the Orchidaceae, a plant family known for its garish displays?

## A WORD ABOUT CLASSIFICATION OF AUSTRALIAN ORCHIDS

Before I answer that question, we must pause for a moment to consider their scientific names. There was a massive revision of the classification of Australian orchids with the publication of the third edition of *A Complete Guide to Native Orchids of Australia* by David Jones in 2021. Among taxonomists, David Jones is known as a splitter and has manufactured many new genera out of old ones. He split the greenhoods into 11 genera, the mayfly and gnat orchids into four, and the helmet orchids into another four. As Jones' treatment is new and has not been well accepted by other orchidologists (including me) I decided to write this chapter using older, accepted names for one important reason. People interested in reading more about these winter flowers will have problems finding them in library books and on the Web using such new names as *Townsonia*, *Anzybas*, *Nematocras*, *Oligochaetochilus*, *Urochilus*. It's enough to say that Jones' book is the most beautiful and informative book on the topic I've ever read. It is unique. How many books feature a full colour photo of a shingleback lizard (*Tiliqua rugosa*) eating a wax lip orchid (*Glossodia major*)?

Dr Jones also published several observations on the pollination of Australian orchids and, for the reader's benefit, I will pay strict attention to his older papers and that means using the older scientific names. They are in the tradition of past generations of those gifted people who went out and studied bush life while better funded scientists in universities and museums were uninterested. During the last decades of the 20th century and the first two decades of the 21st, naturalists paid some attention to a few of the winter orchids I will describe.

## FUNGUS GNATS FAVOUR GNAT ORCHIDS

So how do we explain the seasonal drabness? The best place to begin are the mayfly and gnat orchids. They are small plants producing only one ground-hugging leaf each winter and a single flowering stalk. The construction of their dull-coloured flowers resembles the blunt-leaved orchid in Chapter 3. As usual, there's a stalk-like column combining male and female organs producing pollinia (pollen balls or masses), which may be partially hidden by an overhanging, bonnet-like dorsal sepal. As in the blunt-leaved orchid, these Aussie winter orchids also subdivide their pollinia into two parts so small insects are not overburdened by the weight of all the contents in the solitary, pollen-making stamen as they back out of the flower. As expected, the mayfly orchid's column hangs over and is opposite to the lip petal (labellum), but the lip of a mayfly orchid forms a landing platform not seen in the blunt-leaved orchid. In this flower, the pollinator can perch directly on the lip while it drinks instead of clinging to other floral parts. Unlike the blunt-leaved orchids of Wisconsin bogs, mayfly and gnat orchids lack the hollow, nectar-filled spur at the base of the lip. These Australian flowers are definitely not for

Flower of *Acianthus pusillus* and its fungus gnat pollinator. Note the many pollinia on the gnat's back and on the stigma of the orchid's column. Photo by Rudie Kuiter.

mosquitoes with long, pointy mouthparts. Mayfly and gnat orchids secrete droplets towards the bases of their flattened lip petals. No visitor needs a long proboscis to lick up the reward.

In fact, some Australian naturalists are more likely to notice the mayfly orchid (*Acianthus caudatus*) because it is also known as the dead horse orchid. That should give you an idea of how it smells. David Jones published a paper on its pollination in 1974 and complained of its obnoxious odour, detectable when the daily temperature was only 18°C. Jones felt the stink was strongest just before spring rain. Which pollinators patronise these flowers under gloomy skies? The only insects Jones saw backing out of the dead horse flower wearing pollinia on their backs were fungus gnats in the genus *Mycomya* in the family Mycetophilidae. More recent field studies by Rudie Kuiter show that the orchid is widespread and, in isolated populations, three or more different species of fungus gnats carry whole pollinaria. In any case, these insects are only a few millimetres in length, so they are as easy to miss as the little flowers they frequent.

There's a problem here with these orchids and the dimensions of their pollinators. Like most of their kind, fungus gnats have long, jointed and dangling legs. Propped up on such long limbs they should not fit between the pollinia-filled column and the flat lip petal. Rudie Kuiter remains the most talented photographer of orchid pollinators in Australia. He has taken a series of shots of these insects drinking nectar in the flowers of mayfly and gnat orchids. Remember, unlike mosquitoes, fungus gnats lack elongated mouthparts to suck up nectar. They must lower their bodies until their thoraces touch the lip petal and they can push their immoveable heads directly into the nectar droplets at the base of the lip. I would like to say they get down on their knees but as their leg joints hinge in an opposite direction to our own that comparison doesn't quite work. Once the fungus gnat 'lies down' on the lip petal, the bulge on the back of its thorax becomes positioned directly under the two glue guns made by the special stigma lobe (rostellum) in the overhanging column. You can guess what happens next as the satisfied insect begins to stand up and back out of the flower.

Rudie's ongoing field studies and photography in rural Victoria indicate that different mayfly and gnat orchids can bloom together in the same woodland without competing for the same pollinator. Fungus gnats, in the family Sciaridae, are much more likely to carry the pollinia of the small mosquito orchid (*Acianthus pusillus*). Meanwhile, the common gnat orchid (*Cyrtostylis reniformis*) depends on fungus gnats in the genus *Sciophila* belonging to the family Mycetophilidae.

While fungus gnats occur around the world, they all navigate best in environments filled with organic debris or rotting vegetation at rather low light levels. In such dim and brownish habitats they need not become impressed with vivid colours, with a few exceptions. They are such a diverse group of small-bodied flies requiring mushrooms to complete their lifecycles that scientists segregate these species into at least six different families. Yes, some lay their eggs directly on the plump, fleshy bodies of mature mushrooms and we true gastronomes hate to find a tasty portobello or wild bolete ruined by equally gluttonous maggots. In more cases though, the female gnat lays her eggs in the grey and brown detritus on the forest floor, often burying herself to do so. Some species push their eggs into fungus-infested wood. Once the slimy, tubular maggots hatch they burrow down in search of mycelia. You know mycelia as the feeding webs made of skeins of fungal threads that consume the remains of organic materials. Some will unite and spin themselves into a fresh generation of mushrooms, toadstools and puffballs. The maggots, then, feed on the threads that are in turn, feeding on fallen leaves, twigs and decaying logs. In the temperate regions of southern Australia mycelia grow well in the cool winter when the forest floor stays wet for days and winged gnats look for prospective nurseries, untroubled by lower temperatures.

At the commercial level though, plagues of fungus gnats can devastate mushroom farms as these insects may also carry alien spores that infect the growth medium and introduce diseases that ruin the button stages rising out of the compost. What surprised me was learning that fungus gnats are also despised by people who have never grown a mushroom or collected wild ones in season. Fungus gnat haters include people obsessed with potted plants on their windowsills and in their home greenhouse. Much of what we call, potting medium is based on aged and processed peat mosses. Media used to grow tree-dwelling orchids and other epiphytes are usually made of wood chips. They both become spore-nurturing sinks when kept wet. The spores geminate, fungal threads emerge and attract egg-laying gnats. While fungus gnats are relatively harmless to healthy plants, most home gardeners don't want to be met with dancing swarms as they water their pots.

Kill the maggots and trap the adults is the cry of do-it-yourself-horticulturists with YouTube channels. One video features a woman commenting on the relative advantages of six different anti-gnat poisons on the market. One man had some excellent micro-footage showing what happened when he hit maggots with different chemical solutions. All authorities appear to agree that the winged adults are attracted to yellow objects. Yes, there does appear to be one bright colour to which some fungus gnats will visit in flocks. The advice of the potted plant experts is to find a yellow plastic or paper plate. Smear it with petroleum jelly, stake the plate in a pot and wait as landing gnats become trapped.

As long as there is moist and rotting vegetation full of mycelia, fungus gnats are here to stay. Their fossils dot the interior of Baltic amber. They were probably pollinating orchids with small flowers at that time, as one 45–55-million-year-old specimen died carrying orchid pollinia under its legs. This is exciting as several tropical American *Bradysia* gnats (Sciaridae) continue to pollinate flowers of tree-dwelling *Pleurothallis* orchids transporting the flower's pollinia on the same parts of their bodies. Fungus gnat pollination isn't confined to orchids either. All over the world, the same insects need the energy in the sugars of flower nectar to lay their eggs in moist layers of debris accumulating under trees. A 2018 paper written by two scientists in Japan, Ko Mochizuki and Atsushi Kaeakita, found fungus gnats actively visiting flowers in spring forests and along mountain streams. The gnats drank the nectar and carried the loose pollen grains of wildflowers like miterworts *(Mitella)* and twisted stalks (*Streptopus*), but they also liked blossoms of woody plants including spotted laurels (*Aucuba japonica*), spindle trees (*Euonymus*) and the redbud hazel (*Disanthus cercidifolius*). Most of the fungus gnats checked for pollen were females. Like the winter orchids of Australia, the Japanese plants attracting fungus gnats were shade tolerant and had drab, rusty-red, maroon or greenish flowers.

## GREENHOOD ORCHIDS, PUNISHING SEX AND THE MALE GNAT

Which takes us back to Australia and a look at the greenhood orchids (*Pterostylis*), which are far more diverse than the mayfly and gnat orchids discussed above. There are only 20–21 species of mayfly and gnat orchids, but there are almost 300 species of greenhoods in Australia. None of them make nectar. We like to think that cheaters never prosper, but greenhoods in bloom do nothing but cheat and they are selective fakers. They appeal only to male fungus gnats, the little fools, all lured into the flower by promises of sex. One of the probable reasons why there are so many greenhood species is that their flowers appeal to different species of fungus gnats in more than one family, and the same orchid may depend on different gnats if it grows in different locations and has a broad distribution. For example, the blunt greenhood (*Pterostylis curta*) I examined at Royal National Park, NSW, in the 1990s is pollinated by species of *Mycomya*, also in the Mycetophilidae. In contrast, Rudie studied the same blunt greenhood in southern populations in Victoria. His photos and collections showed that these flowers are pollinated by males belonging to a species of *Mycetophilia.*

With so many greenhood species in Australia, they exploit a far greater diversity of male insects compared to the ones that visit mayfly orchids for legitimate nectar. In some cases, the physical size of the gnat must parallel the size of the flower. The aptly named cobra or superb greenhood (*P. grandiflora*) lures at least two of the larger species of *Mycetophila*. The dwarf snail orchid (*P. nana*) offers a daintier flower receiving smaller gnats that belong to the family Sciaridae. There are other options. Dr Noushka Reiter and her team at the Royal Botanic Gardens Victoria at Cranbourne performed experiments on six greenhood species representing three different branches or lineages in the genus *Pterostylis*. Four of the orchid species seduced four different species of mycetophilid males, as usual. However, the baggy britches (*P. boormanii*) and basalt rustyhood (*P. basaltica*) were preferred by two species in the genus *Xenoplatyura*, and these gnats belonged to a third family, the Keroplatidae.

Greenhood flowers may not look beautiful, but they are whimsical. When the great British botanist, Robert Brown (1773–1858) visited Western Australia in 1801, as the naturalist on the ship HMS *Investigator*, he found and named the first *Pterostylis* species. The name refers to the broad wings on the neck or style of the column. Brown liked to call the flowers druid's caps. I challenge anyone to look at these orchids and not see remnants of the toys or ornaments of childhood. One is reminiscent of the gaping jaws and antlers of a Chinese dragon. Another forms a parrot's head or holds the profile of a gazelle figurine. Others suggest utensils in the kitchen, of an expensive dolls' house, or the helmet of a wind-up robot.

It's hard to look at a greenhood flower and realise it's made up of the same organs as the mayfly, lady's slipper, midge or blunt-leaved orchids seen in previous chapters. Once again, the segments reflect a revisionist streak in floral architecture. The two lower sepals fuse together at their base. When they stand upright, they give the flower its fangs, horns or antennae. Some species do not allow these sepals to stand erect. They dangle down like fish fins, tentacles or ribbons.

Now look at the top of the flower. The curving column of sexual organs is hidden under a dome (galea) made by overlapping the dorsal sepal with the two lateral petals. In some species, this dome narrows to a common point giving the top of the flower a beak, snout or magic wand. In 13 species, the dome elongates into an exaggerated crest and you can see a druid's cowl or, at least, Gandalf's hat. In most greenhoods, regardless of species, the dome is so lightly tinted you can see through the white or emerald streaks and view the column underneath. I saw my first fungus gnat, trapped in the flower, by looking through this dome.

Now take a look at the lip petals. They don't resemble any of the previous orchid lips I've described. Some greenhood lips are broad, some are narrow, and they range in colour from green or brown, to yellow. The sausage greenhood (*P. allantoideum*) has a fleshy, brown lip covered in miniscule warts with a tasselled crest at its base. The lip of the needlepoint rusty hood (*P. aciculiformis*) is smooth, flat and broad, but it has stiff, white bristles along its margins. In the most exaggerated case, the tip of the lips of the 20 species of beard or bird greenhoods are so small, blackish and folded they look like cloven hooves, yet their narrow and elongated lip stalks are densely feathered like ostrich plumes. The species doesn't matter. All of these decorations I've described are dummy females. Their form is supposed to lure and excite incoming male gnats. Most greenhoods also add one more trick that exploits the male gnat's passion.

The lip petal of most greenhood orchid is usually a sensitive organ. It does not flutter uncontrollably in the breeze like the lips of the midge orchids in Chapter 4. A greenhood lip has a joint at its base like an elbow. Just as the two valves of a leaf of a Venus flytrap shut together when touched, the lip petal of a greenhood moves automatically when a little pressure is applied to its surface, and it always springs up. When an insect, in this case a gnat, lands on the lip, the joint receives an electrochemical signal and catapults upwards, tossing the male gnat back into the translucent and hollow dome made by the dorsal sepal and lateral petals. The lovesick insect is now incarcerated in the flower. As in many prisons for humans, there are windows and he may flutter impotently under them in response to daylight passing though the translucent walls of his cell. An additional two broad wings (one on either side of the orchid's column) contribute to the prison's walls and ceiling. If you wish to watch the act of imprisonment yourself, but lack access to native bush in winter or early spring there are several videos online. The easiest way is to search YouTube for '*Pterostylis*' and 'pollination'.

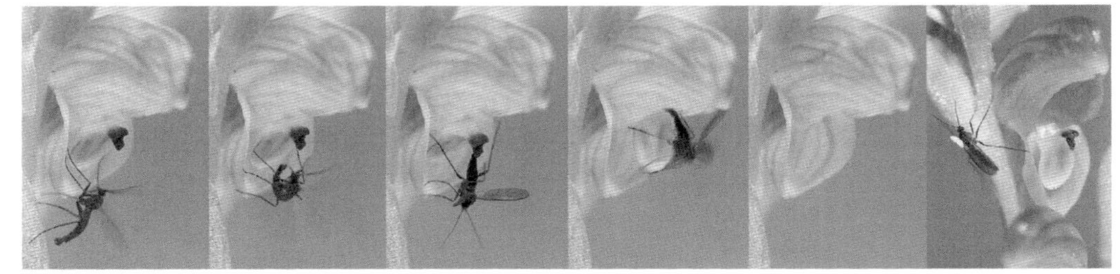

The sequence shows the attempt of a male fungus gnat to copulate with the dark crest on the lip petal of a greenhood flower until it triggers the spring mechanism and is thrown into the chamber made by the hood (galea) while the flat surface of the lip slams the door shut. Once the flower resets there is sufficient space for the gnat to squeeze out between the column and lip petal but the column glues one of two pollinia to his back as a parting gift. Photograph by Tobias Hayashi. From Hayashi T (2022) Sexual deception of male *Bradysia* (Diptera: Sciaridae) by floral odour and morphological cues in *Pterostylis* (Orchidaceae). *Botanical Journal of the Linnean Society* **200**(3), by permission of Oxford University Press.

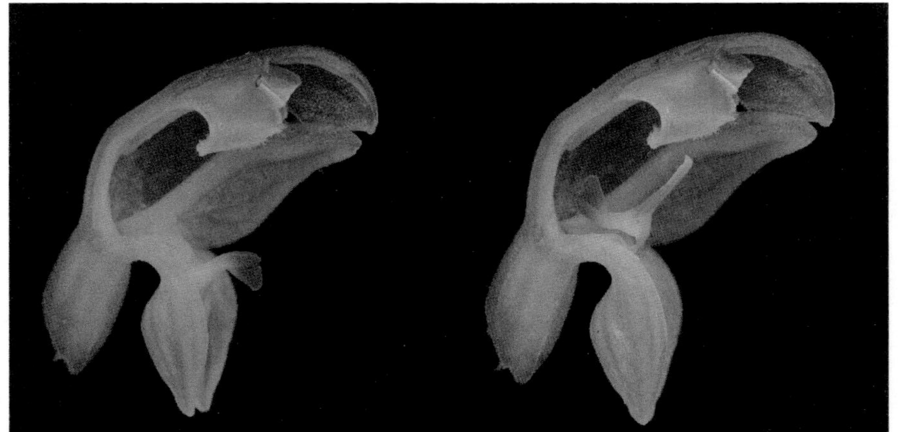

Partially dissected greenhood flowers show how the trap works. (Left) The flower is 'locked and loaded' as the lip petal hangs down exposing the stalked crest serving as the lure to male gnats. (Right) The trap is sprung with the crest thrown backward while the flat tip of the lip petal now rises up to meet the fringed and winged column shutting the 'door' and turning the interior of the flower into a temporary prison. Photograph by Tobias Hayashi. From Hayashi T (2022) Sexual deception of male *Bradysia* (Diptera: Sciaridae) by floral odour and morphological cues in *Pterostylis* (Orchidaceae). *Botanical Journal of the Linnean Society* **200**(3), by permission of Oxford University Press.

To return to the prisoner, the lip petal is now the closed door of the cell but it will unlock and swing open in time. Greenhoods parole their pollinators. Once the lip begins to re-cock and return to its original position on the floor of the flower's cell there will be a slight opening, a sinus, through which the male gnat must escape. This exit forces him to crawl forward onto the lip petal and under the column where his back will contact one of two glue guns located in between the staminode wings. If he is strong enough he will draw out one of the two masses of pollinia on his way to freedom. He then flies away with a yellow pollinium on the upper side of his thorax. Eventually, the lip petal resumes its original position. Now it can wait for the arrival of a second gnat already bearing a pollinarium. This time the spring trap and temporary lock up will lead to cross-pollination and fertilisation of the ovules in the ovary. In this great escape movie, there's no need for a prisoner-dug tunnel, or a file inside a cake.

Of course, a safe exit is not always a guarantee and some prisoner gnats suffer lingering deaths. The first time I found a gnat in a blunt greenhood in Royal National Park he was a corpse. It's not too unusual to find dead insects inside an orchid that makes a chamber flower. Some visitors are not legitimate pollinators and they lack the wit and/or strength to find the exit inside the complicated architecture. It's the same unhappy fate for some insects that entered my mountain lady's slippers (Chapter 2). As usual, it's hard to give the exact cause of death of any insect that didn't find its way out of a chamber orchid. Was it injury, exposure, starvation or merely old age?

Rudie has sent me several photos of fungus gnats that came to untimely ends in flowers of striped (*P. alata*) and scarlet (*P. coccina*) greenhoods. They were too weak compared to the robust construction of the flowers. Some lacked the strength to yank the pollinia all the way out of the anther after their backs stuck to one the glue guns on the stigma. The gnat dangled from the orchid's column until he expired.

## MALE GNAT MORTALITY, SOME UNANSWERED QUESTIONS

Other gnats escaped with pollinia on their backs, as usual, but then fell victim the second time they entered another flower of the same species. At least, they didn't die until *afte*r they delivered the pollinia to the stigma of a flower on a second plant. You might say they became involuntary martyrs to the act of cross-pollination. Surely, this doesn't happen too often. Under normal circumstances, the pollinium of a greenhood orchid is so brittle or powdery that it fragments or erodes once it contacts the sticky, pollen-receiving surface of the column's stigma. As pollen breaks off and sticks to the stigma, the gnat should detach easily from the adhesive surface of the flower's column. Once the lip petal re-cocks, the gnat escapes a second time although he still carries the deteriorating remains of the first pollinium he received from the first flower he visited.

What happens if the pollinium doesn't fragment and remains whole? Well, Rudie thinks that the longer a pollinium remains on a gnat's back the harder and crustier it becomes as it dries in the open air. A pollinium that fails to crumble upon contact becomes wedged into the stigma. A gnat who lacks the strength to break free dies stuck to the orchid's column. Ironically, while his own life ends the grains in the pollinium are producing tubes on the stigma surface that penetrate the pistil releasing their sperm into ovules. Death takes the pollinator while the next generation of orchids is made.

Some victims are innocent tourists entering a flower they can never leave. They do not belong to the right insect species that best fits the architecture of a specific greenhood flower. Surprisingly, some are fatally inquisitive females. However, as males and females of most fungus gnat species are of two different sizes (see below) the females never quite fit the flower's dimensions and they come to bad ends. Sometimes the glue destined for the gnat's back, made by the stigma lobe (rostellum), doesn't dry fast enough. The bumbling gnat struggles too much in the wrong direction, bumps into one of the flower's two lateral petals and the glue fixes the insect to the flower for the rest of its brief life.

Someday it may be possible to quantify, compare and contrast the number of 'deaths by pollination' among different greenhood species. If we can, it may have something important to tell us about what we call the evolution of adaptations. There are several questions requiring answers. Does the reproductive success of an individual greenhood plant increase or decrease if it consistently murders its pollinator? What about the range of different gnat species and sizes that failed to escape the trap while wearing the orchid's pollinia? Does this reflect genetic flaws in the flower's mechanics or are we merely observing the consequences of natural weaknesses in insect strength and perception? These gnats don't live long. Perhaps the most common victims in these flowers are usually the senile and feeble of body. If it's 'the orchid's fault' should we blame the genes it inherited from its parents (nature) or do failures reflect changes in seasonal conditions (nurture)? Perhaps flowers of certain greenhood species slay their pollinators because they did not develop properly in the bud due to an earlier season of insufficient water, minerals or sunlight. There is another possibility. Do some orchid populations grow in areas in which the sturdiest and most enthusiastic gnats are now rare and the flowers must depend on remaining species that are puny dawdlers?

Is it really such a terrible thing if a flower is pollinated but the only pollinator it attracted perished when the pollinia he carried nailed him permanently to the stigma? The little woodland tragedy still ended in a seed capsule. Isn't that what fitness means ultimately in evolution? It's measured by the line of generations you've mothered. Ah, but let us remember, these orchid flowers are bisexual and reproductive success must also be measured by the number of seeds you sire. Making lots of pollen grains is a lot cheaper and requires less time than making and developing seeds inside a protective fruit wall. While the flower of each midge orchid in the preceding chapter always releases all of its pollen as one unit at the same time, a greenhood flower can do it twice, releasing its pollen masses one at a time like the blunt-leaved orchid in Chapter 3. One of its stigma lobes, known as the rostellum, makes two sticky plugs (viscidia) and each plug becomes attached to one of the two pollen masses in the male anther. Given the opportunity of lots of gullible male gnats, each greenhood flower can release pollinia twice. However, if pollinia refuse to crumble or smear when they contact a stigma then each pollinium that attaches to a doomed gnat will father seeds in only one flower provided that gnat has the energy and dim-witted interest to consider visiting more than two fake females. For a population of greenhoods, to achieve reproductive success by setting capsules containing healthy seeds every year, the local gnat population must represent a vulnerable yet sustainable resource each flowering season. If the abundance of gnats decline in winter natural selection would favour a greenhood flower that allows their pollinators to carry its pollinia to more than one flower on more than one plant.

## ATTEMPTED COPULATION FOLLOWED BY IMPRISONMENT

Successful or not, the more information I had about this mode of orchid reproduction the more I found the entire process reminiscent of some of the prints in Goya's *Los Caprichos* and any of those old novels in which courtesans detained, humiliated and destroyed their clients' fortunes and reputations. After such a bad experience, one wonders how some male gnats can behave like desperate johns so stupid that they visit flowers of the same species more than once affecting cross-pollination while receiving callous treatment without fathering a new generation of maggots? Dr Daniela Scaccabarozzi and her team filmed the behaviour of male gnats visiting flowers of the red-banded greenhood (*P. sanguinea*) at Western Australian Botanic Garden in Perth, Western Australia. (You can watch it on YouTube.) It's abundantly clear that the males regarded the tassel on the lip as a receptive female. You can see them poking at the tassel with their curled-up abdomens, as that is where male insects carry their penises. And yes, you can see a male struggling against the translucent wall in the dome as if he was shaking the bars of a jail cell window.

The cinematography of passionate fungus gnats forms its own oeuvre. A team of four Australians published the results of their studies on the pollination of the swan greenhood (*Pterostylis cycnocephala*) in 2022. We don't know how many species of fungus gnats are fooled by this flower but all appear to belong to the genus *Bradysia* in the family Sciaridae. Botanist/photographer Tobias Hayashi filmed the process in one flower. The gnat curled and pointed the tip of his abdomen like a gun and used his genitals to grab wrinkles on the blackish attachment on the lip petal that resembled a receptive female. After the insect spent some energy on a few lusty thrusts, the lip petal was triggered as usual, flinging the gnat into the prison. However, in this case, his abdomen remained firmly clasped to the black fake female after the lip petal shut the prison door.

The same team of scientists did some field experiments and came to the conclusion that the seductive powers of the swan greenhood balanced smell and sight. Male gnats continued to find these flowers within 10 min after the orchids were concealed under a porous paper screen. The gnats perched expectantly on the paper. If the fake female on the lip petal was removed the flower's smell continued to attract males. In the absence of the fake female, the males still landed on the flower, performed their usual courtship rituals (fanning their wings, twisting their bodies 180 degrees, curling and pointing their abdomens). But these activities declined quickly compared to the males who found orchid flowers with fake females intact.

How long do greenhoods imprison their lovers and how often does the same flower play the same tricks over one season? That may depend on ambient temperatures and how much sunlight the plants receive as they stand in and out of light gaps on the woodland floor each day. I did a little work on this in the early-1990s on blunt greenhoods. They bloomed in July and I followed the progress of their flowers for a fortnight. After catching two male gnats wearing the pollinia I decided to tickle the lip petals of these orchids to see how long they'd stay shut and whether they'd snap a second time. Under natural conditions, and daily temperatures usually less than 18°C, it took flowers 9–24 h for the lip to reset in full on cloudy, winter days.

Would the same thing happen if I could study the same species blooming under controlled conditions? I was in luck. Back in Saint Louis, I learned that the keeper of orchids at the Missouri Botanical Garden was growing several pots of the same greenhood species. They were all kept on the same greenhouse shelf, under the same hours of fluorescent lighting, on the same cycle of temperatures between 19 and 22°C. I could manipulate 22 flowering stems, and I did so 63 times. The only difference was that the orchids now responded to the seasonal cycles of the Northern Hemisphere. They flowered from late January until mid-March instead of from July to August.

I labelled the potted flowering stems with jeweller's tags. When a bud opened completely and the lip petal was at the correct angle, I tapped it with a probe one to three times to make it spring up into the column. Then I waited for it to reset. Would the mechanism in the potted orchids reset faster or slower than the populations at Royal National Park? Would the lip of a potted flower respond as fast to another probe the second time around, once it reset?

The answer to both questions was a carefully qualified ... yes. A surprising number of flowers reset within an hour, but only 10 flowers, based on 107 attempts made over 6 weeks, also snapped up their lip petals a second time when tapped. Most reset 2–3 h after the initial triggering, but how many could be fooled a second time? Greenhouse pampering made these flowers more excitable. If it took more than 2 h for the flower to reset the first time then 60% of the flowers would trigger when touched a second time on the same day. Of course, in a carefully regulated environment that excluded fungus gnats, my experiments didn't make the flowers set fruit. You could trigger the same flower over several days but it would never pollinate itself. As cited previously, Darwin was right when he wrote, 'Nature abhors perpetual self-fertilisation'.

## BEARDED GREENHOODS AND DAGGER FLIES

We need to backtrack before we pass on to the last group of winter orchids. It looks like the 20 species of bird and bearded greenhoods (see above) have abandoned seduction of fungus gnats. Some bearded greenhoods extend their flowering into mid-spring and on into the first days of summer. They are waiting for larger dupes based on the observations Rudie and his group made on the bearded greenhood (*P. plumosus*) in Victoria. Have you ever heard of the dance, balloon or dagger flies (Empididae)? Amber fossils show they've been around since the Cretaceous. While these insects take nectar from flowers, and some eat pollen, most have a vicious, hooked proboscis built for stabbing and killing other winged insects. When the male selects a female, he presents her with a dead insect. Entomologists refer to this as a nuptial gift but you may prefer to use the equivalent French phrase of *un petit cadeau*, or the Australian, *bonza pressie*. If the female accepts the gift, she will hold it with her legs and eat it while allowing the male to mate with her.

As the female dagger fly expects the male to give her something more nourishing than a box of chocolates (she has eggs to provision) she must compete for her dinner with other females. Some females have evolved feathery scales on their legs and inflated sacs on their abdomens. Both are supposed to make a female more attractive to prospective suitors. The bearded greenhood exploits these feminine fashions. The yellow beard is made of many bristles we presume mimic the scaly legs of female dagger flies, but the orchid exploits the male's lust even further offering dozens of lovely fake legs instead of the usual six. As in the case of other sex mimicking orchids we will encounter in Chapter 7, it looks like the bearded greenhood releases a fake hormone (kairomones) to signal its receptivity to male dagger flies.

Rudie's photo series showed what happened next. Males flew to the orchid often bearing dead insects as large as they were. The edible gift was presented to the dark, cloven hoof at the tip of the lip petal. The male fly must interpret this sculpture as the head of the female. He even stashes the prey between the closest fake legs. Does he have sex with the lip petal? Well, he certainly tries. The photos show how he crawls up the exaggerated lip poking his abdomen under the fake legs looking for a genital pore, but there's nothing there. He keeps ascending the pole of lovely legs drawing closer to the entrance of the flower with its sexual column hidden in its crested dome.

The lip petal of the bearded orchid doesn't jolt upwards suddenly, unlike the fungus gnat-pollinated greenhoods. The dagger fly is not catapulted into the flower's dome. The flower operates in a discrete and subtle manner gradually raising its lip petal as the male climbs and probes the staircase of fake legs. By the time he is lured into the chamber made by the dome, the lip has become a drawbridge shutting the flower's entrance. It probably won't imprison the fly for hours as in other greenhoods. There is a separate, smaller escape hatch under the pointed beak of the dome. To reach it, the male dagger fly passes upwards through the column and the pollinia is transferred onto the back of his thorax.

## FUNGUS GNATS MEET THE PRIESTS OF RHEA

Let's return to the fungus gnats to understand our final group of winter orchids. Curiously, they were named after characters in one of the odder Greco-Roman myths, although our story begins again in southern Australia. Like so many men of his time of a certain class, the botanist Richard Anthony Salisbury (1761–1829) had a classical education. Presented with specimens of what was obviously a new genus of orchids, the flowers must have reminded him of helmets, so he named the genus *Corybas*.

In Greek Mythology, Kronos, King of the Titans, feared one of his children would dethrone him. Every time Rhea (Kronos' wife) presented him with a new baby, he swallowed it whole. While Rhea substituted a stone for her last baby, Zeus, she realised that a crying infant would attract the attention of her insatiable husband. Handing the baby over to some nymphs, Rhea made a race of helmeted warrior priests out of raindrops. When little Zeus cried the men banged on their helmets and shields to drown out the noise. 'That's how my priests worship me,' Rhea explained to her cannibalistic mate. 'Must they be so loud?' he replied.

The dorsal sepal of a flower in the genus *Corybas* forms a broad, curved and inflated dome (another galea) over a good part of the lip petal. In some species, the lateral sepals and lateral petals are long, narrow and curve upwards to resemble the elaborate plumes or hackles on a helmet. In southern Australia, helmet orchids often show one more trait in common with their mythical namesakes. Their flower buds emerge and pop open with the onset of winter rains.

*Corybas* species are distributed from sub-Antarctic Macquarie Island into Tasmania, then north through New Zealand, coastal Australia, New Guinea, into tropical south-east Asia and a few in the lesser Himalayas. While over a hundred species are recorded, they are receiving more attention from tropical botanists who have learned to pay attention to small plants in wet shade. As exploratory botanists learned that helmet orchids hid their charms more species were discovered and described. Some bloom while partially concealed under the leaf litter or obscured by thick moss beds as their flower stalks are so short. The only way you know they are there is that, once again, their lateral petals and sepals are so long and spidery they push up above the litter. These organs are probably wearing the flower's scent glands alerting gnats to their presence.

Until recently, almost everything we knew about reproduction in helmet orchids came from a few studies in southern Australia. Interest began again, with observations by David Jones in a paper published in 1970. Jones dissected a flower of the stately helmet orchid (*Corybas diemenicus*) and found a dead, female fungus gnat carrying the pollinarium on her back. This led to speculation that the orchid was pretending to be a mushroom exploiting pregnant but undiscerning females. There's nothing farfetched about this. Unrelated plants do it all the time. Some of you may own a carrion flower plant (*Stapelia* species). When this desert succulent in the frangipani family (Apocynaceae) blooms, the flower dresses in dark reds and browns, bears tufts of hairs, and stinks like an abandoned carcass. You may observe it is visited by female blowflies that carry off the pollinia but leave their eggs. Of course, the maggots die of starvation after hatching, as the flower can't provide the appropriate fats and proteins. Likewise, traps for egg-laying flies are common in many members of the philodendron or taro family (Araceae). The urn- or vase-shaped flowering stems may smell of mildew, dung or decay, whatever the preferred fly would use as its normal nursery. The only problem with Jones' hypothesis was the lack of a smoking gun. The fungus gnat left no eggs in the flower.

Nevertheless, naturalists have looked into helmet orchids for decades and interpreted them as flowers dressed as mushrooms. After all, the dorsal sepal forms a convincing cap and the margins of the lip petal in many species are scalloped like mushroom gills. Rudie Kuiter and Mitchell J Findlater-Smith finally had enough of these speculations and decided to test the mushroom brood site hypothesis. They found populations of five species of helmet orchids in Victoria and studied them over several seasons.

Several interesting and unexpected discoveries were made. Yes, all were pollinated by fungus gnats but these insects were not early risers. Pollinators began to arrive around lunchtime and their visits would last 1–5 h. Most of the activity occurred on atypically warm, winter days. Different fungus gnats belonging to different species in different genera preferred different helmet orchids. One *Phthinia* species carried the pollinia of the cradle orchid (*Corybas aconitiflorus*), while a second and distinct *Phthinia* species was loyal to the pelican orchid (*C. unguiculatus*). A species of *Mycetophila* visited the slaty helmet orchid (*C. incurvus*). As for the stately helmet orchid, studied previously by Jones, Rudie and Mitchell took another look and found it was pollinated by a new species of *Phthinia*. Did they ever find eggs in the flower? Yes, but they were laid by springtails (Collembola) and some unrelated flies. No fungus gnat maggots were ever found in any of these helmet orchids. Kuiter and Findlater-Smith followed the flying female gnats to rotting wood or bits of fallen bark. That's where the gnats laid their eggs and their maggots matured. The two men even collected the infested debris, brought it back to a terrarium and watched them metamorphose into the appropriate species of fungus gnat.

Male and female fungus gnats aren't the same size and this led to another important series of observations. In some cases, males are bigger than females and in other species the sizes are reversed. This makes a big difference as the distance between the orchid's column and the lip petal is set. The cradle and the stately helmet orchids were both pollinated by female fungus gnats. The slaty helmet and the fringed helmet orchid (*C. fimbriatus*) were dependent on males. The flowering of the slaty helmet was the site of an orgy at one location. While only males wore the pollinaria, females also arrived and the males gave chase even though the males were already wearing heavy packets of pollinia on their backs.

If different species of helmet orchids bloom together, they should avoid competition because they depend on different species of fungus gnats representing different sexes, but mistakes occur. A new Australian helmet species, *C. miscella*, was discovered and named by Jones in 1991. He has since admitted his error. The plant is a recurrent hybrid between the slaty helmet and the closely related veined helmet orchid (*C. dilatata*). As hybrids go, it's quite vigorous and may form large vegetative colonies of interconnected clones among the fallen pine needles in pine tree plantations. Jones has also reclassified a second hybrid (*Corybas* x *dowlingii*) once known as red lanterns, and it is now treated as a hybrid between the cradle orchid and the fairy lantern (*C. barbarae*).

This takes us back to the question. Why did fungus gnats of either sex visit these flowers in the first place? Rudie made a strong point in emails to me and followed it up with photos. Specifically, helmet orchids have something on their lip petals that does not resemble a mushroom. It is a smooth, rounded bump or boss. That boss is usually a contrasting colour to the margins of the lip petal and the helmet-shaped dorsal sepal. It's a white or pink bullseye surrounded by the brownish-iodine pigments on the other flower parts. The boss is where male or female fungus gnats feed. Once again, they lower their legs to take nectar, but this time the liquid reward is secreted by the contrasting boss on the lip petal lying under the flower's always ready column of sexual organs.

## CHINESE HELMET ORCHIDS, MAKING SENSE OF SCENTS

Additional studies on the pollination of helmet orchids in Malaysia and New Zealand were published but without much supporting evidence of fungus gnats in action. I was involved in a recent study in Yunnan by members of the Kunming Institute of Botany. As I could not go to China in 2020 thanks to international travel restrictions due to COVID, all I could do was sit on the sidelines, thousands of kilometres away suggesting techniques and equipment to collect data. The team at the Institute turned to two helmet orchid species found on lower subtropical slopes and valleys in south-western Yunnan. I certainly missed a glorious trip as one species, at 1,300–1,400 m a.s.l., lived under broad-leaf subtropical forest with a canopy dominated by needlewood trees (*Schima wallichii*) and Chinese coral berries (*Ardisia shweliensis*). The second at nearly 1,800 m a.s.l. was a wildflower of pine-oak forest with a glorious understorey of rhododendrons.

You may want to call us mere copycats following Australian successes. I disagree. Yunnan is part of the edge of the natural, northern distribution of all helmet orchids. Comparing aspects of the lifecycles of related species where they are most diverse versus where they are least diverse can be very rewarding, as it teaches us about evolutionary limits to natural distributions. My Chinese co-workers had a couple of things to consider. The two species they studied bloomed in summer, not in the Chinese winter. Were they pollinated by gnats in the same families as the ones Rudie observed and collected in Australia? Perhaps the Chinese orchids were the real mushroom mimics and that was of major interest to Yunnan scientists for one big reason.

Wild mushrooms remain the favourite wild food of the people of Yunnan from mid-summer to early autumn. Families living in small rural villages and farms make a second income collecting dozens of species to sell in city markets and to restaurant distributors. On any day, over 20 native species of edible fungus may be for sale in an open-air market and those are only the ones scheduled for dinner. It does not include other species purchased more for medicine or as table decorations. There's quite an export industry too, with the most desirable fungi sold to the Japanese. Some are processed as mushroom chips or added whole to bottles of alcohol to flavour potent liquors. Mushroom taxonomists in China insist that recent genetic evidence shows that many undescribed species are mistakenly lumped with the popular ones. The mushroom scientists, or mycologists, of China insist their population is composed of millions of gourmets who don't know they are dining on at least 900 species, not 20. The pollination of helmet orchids should receive more attention to better understand the welfare of prized delicacies.

After living in Yunnan for 5 months over 2 years I'd eaten more than my share of chef-prepared fungi, and considered myself most fortunate. It was easy to understand why Yunnan's botanists wanted to understand more about the insects and orchids that could have some impact upon a beloved and sustainable resource. I had an additional motive. Did the flowers of these helmet orchids smell like mushrooms? After all, these flowers bloomed in the same micro-habitats and appeared at the same time as the fruiting bodies of many mushroom species.

Back in the mid-1990s I'd accompanied the Keeper of Orchids at the Royal Botanic Garden Sydney, Dr Peter Weston, to a reservoir. There was a large, mat-like population of one of the smaller helmet orchid species. We collected a few for the collection. I smelled the flowers after they were cleaned of soil and litter fragments. They did smell like mushrooms. To be specific they smelled like commercial button mushrooms that had been placed in a covered pot with a little butter and steamed in their own juices.

Scent biochemists point to at least three signature molecules secreted by ripe mushrooms (1-Octen-3-ol, 2-Octen-1-ol and 3-Octanone). The same chemicals are pumped out by some tropical American orchids in the genus *Dracula*. These *Dracula* flowers are pollinated by female flies that lay their eggs on the orchids' lip petals because the petals mimic the shape and smell of mushroom gills. Could we find these scent notes in either of our two Chinese helmets? Did our orchids make molecules that overlapped with the odours of the slippery jack (*Suillus bovinus*), deceiver (*Laccaria*) and rosy-gill (*Mycena*) mushrooms found at our Yunnan sites? One population of *Corybas geminigibbus* (sorry, there are no common names for these plants in Chinese, as far as the author is aware) shared its site with at least 16 mushroom species according to the specimens identified by members of the mycology department of the Kunming Institute of Botany.

Our team didn't find eggs in the flowers of either species of helmet orchid, although there were unidentified eggs of flies on many resident mushrooms. Flowers of *C. geminigibbus* were visited far less frequently by resident gnats, although these little orchids were surrounded by a great diversity of fruiting mushrooms. Fungus gnats were observed visiting the orchids in bloom only three times after 300 h of observation and they only appeared from 11 a.m. to noon. A female *Phthinia* carried pollinaria on her back as she left one flower. As in southern Australia, score one for gnats in the family Mycetophilidae.

Flowers of *C. shanlinshiensis* were a little more popular and shared their habitat with far fewer mushroom species. After 150 h, three fungus gnats were collected. Score two for the Mycetophilidae, but this time all three belonged to the genus *Exechia*, a novel discovery in the

A female fungus gnat (*Phthinia*) wears the pollinia of *Corybas geminigibbus* on the back of her thorax. Photo by Zhoudong Han.

study of orchid pollination. These new gnats had two periods of flower-visiting activity. They first appeared from 11 a.m.–2 p.m. then they took a break reappearing from 4–5 p.m. A male and a female were found to carry the pollinaria on their backs.

Unlike the observations and photos in the earlier Australian study, neither genus of fungus gnat was observed licking the boss on the lip petal of either orchid species. Dissections of the flowers failed to turn up evidence of liquid secretions from any floral organ. When an *Exechia* gnat entered a flower of *C. shanlinshiensis* it climbed over the boss and crawled down the flower's narrow floral tube. In the absence of drinkable liquids, we don't know what it was looking for or if it found what it wanted. We agreed that both orchid species were poorly visited because they were only faking some sort of reward.

Our two Chinese helmets may survive because they can convince enough native gnats to visit just a few flowers during their respective flowering seasons. But the reproductive success of any faker is extremely dependent on the long-term gullibility of its preferred pollinators. This was reflected clearly in different rates of fruit set by the two helmet orchids in the same year. While fakers spend fewer resources on keeping their pollinators satisfied, their deceived clients are less likely to revisit a second or third flower of the same plant species. When a pollinator learns from its initial mistake, the pollinia they removed from one orchid faker may never end up fertilising the ovary of a flower on a second plant. The frequency of fertile cross-pollinated seed tends to be low in many orchid species with deceptive flowers as we saw in some of the species of lady's slippers in Chapter 2.

Almost a third of the flowers of *C. geminigibbus* set fruit at their sites. That's quite respectable for an orchid species offering flowers that fail to contain any obvious reward. *Corybas shanlinshiensis* had a bad year though. While its flowers seemed more popular with pollinators its natural rate of fruit set was less than 8% and that sends a powerful message to us when we try to conserve remaining populations of rare orchids. In some years, cheaters don't prosper. As neither orchid species self-pollinated after they were isolated under protective bags that excluded potential pollinators, we can say that fungus gnats are essential to seed set in both species.

Well, what about the scent evidence? We had interesting results but they were not definitive. As expected, the three molecules associated with mushroom aromas were made by the local mushrooms, but not by the same mushrooms. The deceiver mushroom made all three. The slippery jack and the rosy-gill made only 1-Octen-3-ol; the flowers of our two helmet orchids lacked all three scent notes. Nevertheless, there were some unexpected overlaps between the mushrooms and the flowers. Both orchids and the slippery jack secreted longifolene. We humans think longifolene produces a woodsy note and it makes up one of two dominant fragrances we smell in a brewed cup of lapsang souchong tea, as the tea leaves are smoked over pinewood fires before they are sold. Both orchids and the deceiver mushroom also released L-alpha-terpineol and this molecule is one of Nature's more common scent notes. You have smelled it in the fragrances of leaves and stems in the mint family (think oregano, rosemary and marjoram), in the flowers of narcissus and freesias, and while grating the zest of lemons.

At first, we were excited when our analysis detected 1-Octanol in the flower of *C. geminigibbus,* and in the deceiver and rosy-gill fungi. We wrote to Professor Robert Raguso of Cornell University who works on scents shared by orchids and mushrooms, and he promptly crushed our hopes. The flower couldn't be mimicking the fungi because the orchid secreted many times the volume of 1-Octanol produced by either of the two mushrooms. Furthermore, 1-Octanol is a molecule lots of other flowers in unrelated families make and then mix with other compounds to produce their characteristic scents. None of the flowers of unrelated plants, including additional orchid species, depended on fungus gnats for pollination.

A little research produced more questions than it answered, a common scenario in science. What we failed to answer will intrigue the next generation of scientists with better funding and new equipment. Nevertheless, I think it's reasonable to retire the idea that the majority of helmet orchids are pretending to be mushrooms, don't you? Female fungus gnats in the family Mycetophilidae, at least, don't want to lay eggs on mature, spore-making mushrooms. They want their offspring to eat the mycelia growing in leaf litter and decaying timber.

My colleagues and I are a bit divided as to the best interpretation of the presentation of the flowers of both species. As these tiny flies like to hide in the litter when they are not feeding or mating, the Chinese scientists think the helmets may serve as temporary shelter flowers. Fungus gnats are such delicate little things and the orchid flower may offer the same protection as a tent made out of a dried and curved leaf or a piece of bark. It's a reasonable assumption as, in other parts of the world, some flowers serve primarily as dormitories for their pollinators. In the Middle East, for example, male bees are the main pollinators of the long-lipped serapias orchid (*Serapias vomeracea*) and the famous black and brick-red *Iris* species. The bees sleep in these flowers overnight and transfer pollen from plant to plant, as they squabble over preferred bedrooms at sundown.

In contrast, I still go for the standard food mimic explanation. The flowers of both species are really quite striking up close. Both display a flamboyant boss on their lip petals surrounded by raspberry-red or banded lip petal margins. As we now know that Asian forests contain other wildflowers and shrubs pollinated by fungus gnats maybe we should compare the helmet orchids to unrelated species in Yunnan that bloom in the shade, have red and/or white petals, but secrete nectar. Come to think about it, we could both be right. Perhaps the helmet orchids of China pretend to offer the equivalent of a bed and breakfast.

It looks like some fungus gnats don't deserve our contempt. Most stay in the wild and don't infest our plants or leave their babies in the commercial mushrooms we prize in our soups and stews. The same can't be said about the insects in the next chapter as we return to the Northern Hemisphere. Like the fungus gnats in this chapter, no one expected that the aggressive cadets of a royal air force pollinated anything.

# 6
# Waspish

*Within the chalice of a flower*
*A bee "improved the shining hour,"*
*Whom, when she saw, a wasp draw near,*
*And sought to gain the fair one's ear,*
*With tender praise: "Oh, sister mine–*
*(For love and trust that name entwine)"*
*But ill it pleased the haughty bee,*
*Who answered proudly: "Sisters!–we?*

– The Wasp and the Bee, *Sarah Anne Curzon*

Did I just see a large wasp fly into that strange, nodding flower? The Grande Ronde Overlook Wildflower Institute Serving Ecological Restoration (GROWISER) in north-eastern Oregon, USA, is large enough to incorporate several overlapping ecosystems. The landscape in which I am working has an unattractive and misleading name – local geologists call it scabland. The

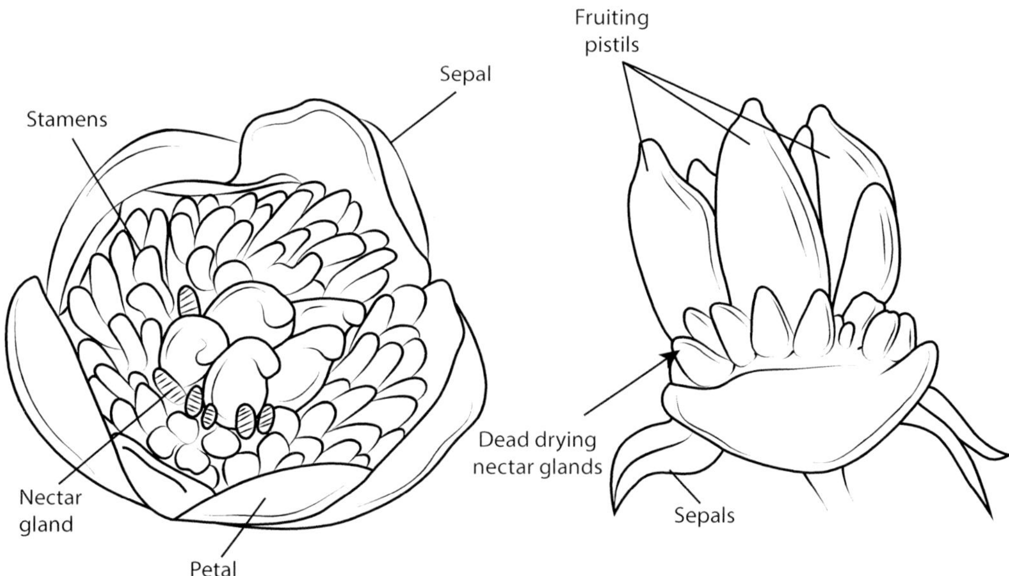


Flowers of Brown's peony (*Paeonia brownii*). (Left) Flower with five pistils surrounded by knobby nectar glands that are almost covered by the many stamens. (Right) As pistils ripen into fruits the petals and stamens drop off exposing the dehydrated nectar glands (arrow). Adapted from photos by Andrew Huber and Nan Vance. See colour section also.

vegetation grows on the rough and thick basalt of an extinct shield volcano. You'd never agree with that classification if you visited the property in late April when it blooms with some of the most iconic plants of the Pacific Northwest. I counted over 20 species including native roses, delphiniums, beardtongues (*Penstemon*), western monkshood (*Aconitum columbianum*), and of course the impressive yellow heads of mules ears (*Wyethia amplexicaulis*) and two species of balsamroots (*Balsamorhiza*).

My attention, though, was drawn repeatedly to an unfamiliar plant with silvery-green foliage. There were over 100 plants and each one formed a clump of two to 10 stems. The tip of each stem ended in one compact flower hanging horizontally or nodding on its stalk. When I held the stem and turned the flower upward, I was surprised. It held its organs tightly and looked like it was stitched out of leather. The green sepals were thick and succulent. The petals were brownish red or burgundy with yellowish tips. A spiral of 84 yellow stamens was packed around five, fat. reddish ovaries each tipped with a stigma resembling a green fingernail. The flower smelled 'bitter', reminiscent of chopped rocket. I tasted the copious nectar and found it sweet but with a bitter aftertaste. Such were my first impressions of Brown's peony (*Paeonia brownii*), and yes, wasps were part of the menagerie visiting the flowers.

## WELL-KNOWN VERSUS LESSER-KNOWN PEONIES

Hybrids and mutations derived from at least a dozen *Paeonia* species out of the 30 native to Eurasia and the Middle East have enjoyed the status of beloved ornamentals for more than a thousand years. Few botanical gardens in China lack tree peonies based on at least three wild species and they have been in cultivation since the 6th or 7th century CE. Chinese tree peonies flaunting red petals are regarded as the King of Flowers symbolising wealth and power. Those offering other colours may represent female beauty, erotic desire and the spring season. Even the starchy roots of more common herbaceous peonies remain part of the traditional Chinese pharmacopeia. They are sliced into round tablets and dried.

In Europe, herbalists addressed peonies with important titles like *officinarum/officinalis*, as they had official uses as pharmaceuticals. Babies were given the root of *Paeonia officinalis* to teethe on and adult Romans were advised to place them under their pillows to avoid nightmares induced by tricky fauns. Peony-based remedies had a long lifespan in Eastern Europe. As late as the first three decades of the 20th century Jewish women of the Pale of Settlement boiled the flowers and drank the potions to treat noisome menstrual discharges. Maybe peonies grown in Europe didn't inspire as much art or poetry as in China or Japan, but they do appear in four centuries of Dutch and Flemish still lifes. They also highlighted some of the canvasses of French artists including Auguste Renoir, Paul Gauguin, Claude Monet and Frédéric Bazille.

Did either of the two species of peony native to America's Pacific Northwest receive as much attention? Yes, there is also a California peony (*P. californica*) blooming further south along hilly coastlines and in chaparral communities consisting of thickets of dwarf shrubs adapted to Mediterranean climates. Some American First Nations people are said to have believed the plant had medicinal properties, and that its leaves and roots were best used to fatten their horses and women. In contrast, pressed plants and viable seeds of Brown's peony were first

Door ornament photographed outside the house of one of the Naxhi people in Yunnan, China. The woody stems suggest that domesticated varieties of native tree peonies were used as models. Photo by Peter Bernhardt.

collected for the botanical gardens at Kew in 1826 by Scotsman David Douglas (1799–1834). He found his first plants in the Blue Mountains of what today is north-eastern Oregon, one of the regions where I worked. The new wildflowers excited him so much he named it for Robert Brown (1773–1858), the most famous Scottish botanist of his day and the man who described the movements of particles in fluid (Brownian motion). Douglas so esteemed this peony he wrote, 'This valuable addition will I trust be an acquisition to the garden ; if in my power, seeds of it must be had.' What was the result? Well, do you grow either American species in your garden? In general, horticulturists regarded these atypical flowers as the only ill-favoured daughters in a family blessed with adorable debutantes. If these American species are grown at all today, it's for the striking colour of their leaves and every good collection should be rounded out with a couple of curiosities.

## THE 'JUICY' FLOWERS OF BROWN'S PEONY

I didn't think of them as things best left to stand alone in a corner and neither did the USDA Forest Service funding the research. I would enjoy the contributions of a consortium of scientists and graduate students who joined field studies and lab analyses over 2 years. This included Dr Nan Vance, my supervisor at the USDA, and Peter Goldblatt, a curator at the Missouri Botanical Garden. Nan insisted we needed a second population to observe and she found one on the eastern flank of the Oregon Cascades Range. You see, I had a problem with time and space. The plants bloomed from April to mid-May. My university was not going to let me walk away from 2 weeks of lectures. The flowering periods of both populations had to be monitored as a single flower of Brown's peony could live from 9 days to 2 weeks. Sending Drs Vance and Goldblatt out to sites with butterfly nets and killing jars was a wise move as we learned there was a changing of the guard of flower visitors over a month. Lots of insects visited the peonies' flowers but not all of them were pollinators. For such a nondescript and modest flower, we learned it had an outstanding way of making itself popular, even though it was surrounded by the most gorgeous yellow, blue, pink and purple competitors.

This was obvious if you found the flowers early in the morning before the insects arrived. In between the last swirl of stamens and the central crown of carpels is a ring of irregular green knobs. Each knob has a little depression that leaks nectar, an awful lot of nectar. In fact, to this day I have never worked on another flower that secreted so much fluid. I had to use the longest glass microcapillary tubes to drain the green knobs. Think of a hollow microcap as about the thickness of the quill in a sparrow's wing feather. Each flower contains a circle of at least 12 green glands. On any given day, three glands secrete nectar. Each gland can ooze an average of 7 μL (microlitres) of fluid but far more is possible. Some flowers produced nearly 32 μL and that's extraordinary when you think that most of the species of wildflowers I've poked into for nectar yielded less than 6 μL on a good day. Every midge orchid flower, described in Chapter 4, produced less than 1 μL.

The volume of nectar offered daily by the flower didn't change much as it aged. The amount of sugar dissolved in a nectar gland's secretion increased a bit over time, growing more concentrated as the blossom aged. The actual concentration of dissolved sugars usually ranged from 18 to 23% over a flower's lifespan. When simple sugars were identified using samples allowed to dry on filter paper the readings for glucose (37%) were more than three times higher than fructose (11%). We even found a dash of maltose in nectar isolated from insects under a bridal veil bag so we know that the flower made it all by itself. It couldn't have come from the saliva of a visiting animal. The dominant sugar appeared to be good old sucrose. It was a pleasantly calorific brew but something more, something we couldn't taste, was present.

Those spots of nectar were mailed to the laboratory of Dr Jiping Zou and the late Dr Gary Booth (1940–2019), both biochemists at Brigham Young University. They gave us the names and concentrations of sugars first. Then we learned that peony nectar was also a cold broth of 16 essential amino acids, the building blocks of protein. That's exciting because you can't make a new generation of pollinators out of mere sugar and water. Amino acids build the proteins in eggs and embryos. There's plenty of amino acids in pollen but, as I mentioned in earlier chapters, many pollinators are unable to access the nutrients in pollen grains.

Small wonder that so many insects visited these flowers each spring, but which of them were regularly transferring pollen from one flower to the blooms of other plants effecting cross-pollination and spreading beneficial genes? Our experiments showed that a flower would accept the pollen of a second flower on the same plant, but that was just another form of self-pollination leading to inbreeding. It was unlikely this self-marriage produced filled, fat seeds that had to overwinter in cracks in the thin soil. Our observations showed that an individual flower attempted to isolate itself from its own sperm by staggering the maturation of its sex organs. The first day the sepals opened, they still clenched the corolla tightly so the petals opened incompletely like pursed lips. This exposed the green stigmas on the ovaries to incoming pollen but the stamens in the same flower remained tightly shut and hidden. For the next 2 days, the flower 'caught' pollen delivered by thirsty insects.

As the flower aged, the sepals and petals relaxed. The receptive stigmas on the flower dried and folded up while the freed stamens cracked open offering hundreds of thousands of grains. For the remainder of the floral lifespan the flower offered its pollen but never accepted any more grains. When female organs in a flower mature first and are isolated temporarily from later maturing stamens in the same flower botanists call this mode of development protogyny.

## VISITORS VERSUS TRUE POLLINATORS

While the peonies were well attended by small, solitary bees (*Andrena* and *Lasioglossum* species) we had to interpret them as robbers, not pollinators. They were happiest taking pollen and nectar during the flower's male phase, but by then the ovaries were too old to receive pollen. These tiny bees had trouble navigating the more constricted female phase. The most spectacular visitor, especially in the first 2 weeks of bloom, was a huge, noisy hover fly, *Criorhina caudata*, at least 1.5 cm in length. It was one of those hover flies that pretend to be bumblebees. It made a dramatic buzz, of course. It also had a banded abdomen and it was very hairy, just like a bumblebee. Naturally, it has no stinger but, presumably, inexperienced birds, lizards and toads don't know that. We caught 18 specimens over the years amounting to 17% of the total insect catch.

Our hover fly was a peony pollinator. As it drank nectar in flowers in the male phase, its face and thorax became dusted with pollen. When it switched to a female phase flower, it shoved its proboscis past the pursed petals touching its face to the receptive stigma tips. After euthanising a fly in a killing jar, I would put it on a glass slide, add some drops of solvent and wash off the

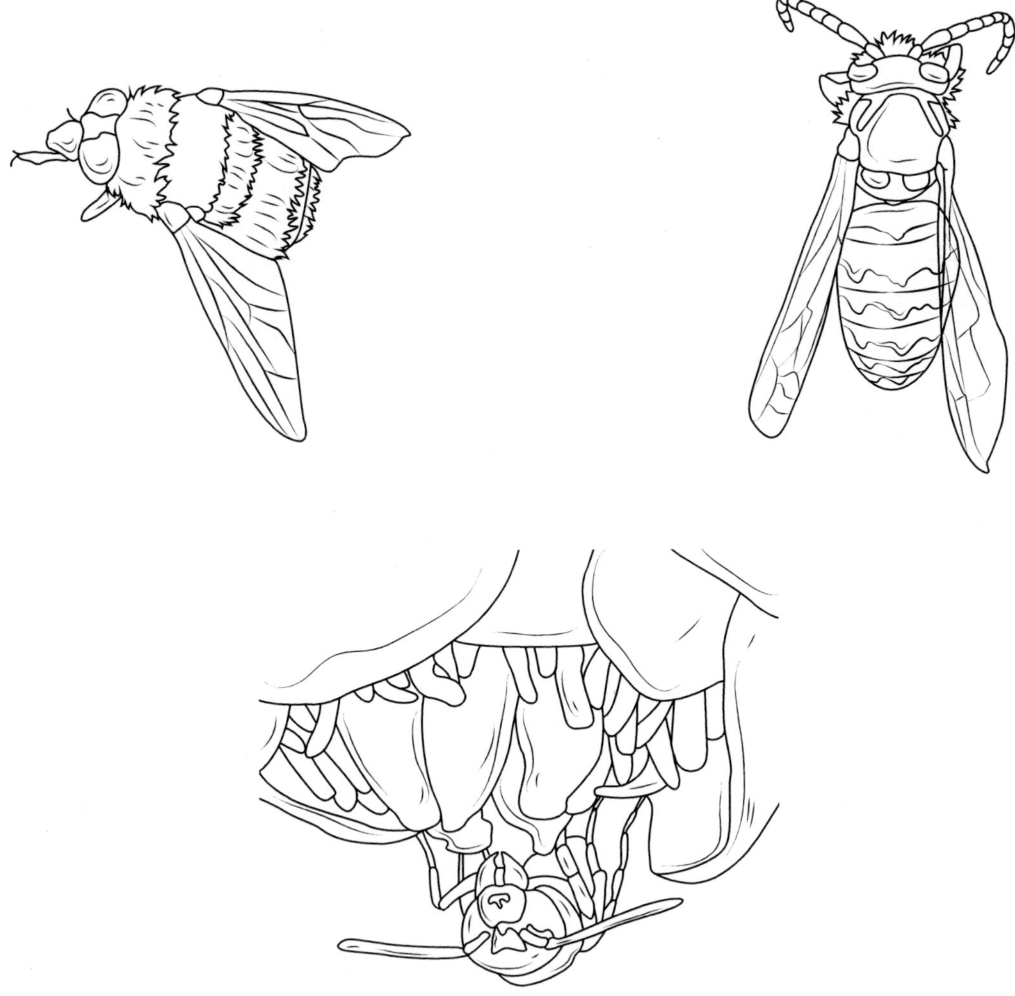

Two of the pollinators of Brown's peony, the large hover fly (*Criorhina caudata*) which mimics bumblebees and a queen yellow hornet (*Dolichovespula arenaria*). Illustration by Envisage Information Technology.

pollen clinging to the hairs. The solvent evaporated and I stained the residue mounting it in a pink dye known as Calberla's fluid. Under a microscope the smear showed that a fly body typically carried more than 200 grains of peony pollen.

Brown's peony stayed in bloom for nearly a month at GROWISER and that fly was fickle. As more wildflowers bloomed as chilly April turned to warm May, the *Criorhina* flies turned their attention to the balsamroots and Nuttall's larkspur (*Delphinium nuttallianum*). Did this mean that all the pollination of Brown's peony occurred within the first week of the season? No. As the attention of the fly waned, a second group of large insects investigated them and they proved to be just as enthusiastic. Over a period of three spring seasons, we observed and collected six species of social wasps on the flowers. Once again, I collected these insects, and removed and stained their pollen. Their bodies weren't as hairy as the fly so it was no surprise that a wasp carried less than half the number of peony grains on her head, legs and thorax. While these wasps were not as efficient as pollen carriers as the flies, they made up for it in numbers and thirstiness. Sometimes as many as three wasps drank the nectar of the same flower at the same time. All wasps visited male- and female- phase peony flowers. With tongues much shorter than the fly's their heads pushed against the pollen-receiving stigmas in female-phase flowers. A slender wasp looked like a lion tamer sticking its head in a big cat's mouth.

## THE NATURE AND NURTURE OF SOCIAL WASPS

All the wasps collected on the peony flowers were social species and female. The most common were yellow hornets (*Dolichovespula arenaria*) and golden paper wasps (*Polistes aurifer*). Seasonal males are not produced until months after the peony blooms. Yellow hornets are found throughout much of North America. The golden paper wasp is primarily an insect of the American and Canadian West. However, both build exposed nests of papery combs made of plant fibres mixed with their saliva. The remaining wasps were American yellow jackets (*Vespula* species). They protect their combs in paper envelopes and usually establish their fortresses underground in cavities like abandoned rodent burrows.

Most of the peony wasps were queens who hibernated over the previous winter. A solitary and fertile yellow hornet or yellow jacket is all that is needed to start a nest in early spring. Once the first generation of workers are winged, a yellow hornet queen stays at home laying more eggs, just like a seasonal bumblebee queen. It's a little different for the golden paper wasp. In late summer, fertile females leave the old nest and may hibernate together over the winter. In March or April, one foundress (her official entomological name) starts building a nest to attract her sisters, now known as subordinates. I found some of these early combs partially exposed between scabland rocks. The subordinates play the roles of ladies in waiting until the foundress lays a first generation of eggs and leaves home for good. The subordinates take over reproduction and their new generation consists primarily of sterile works.

We wandered close to wasp nests while catching them on the peony flowers. Should we have been more cautious? No one was stung over those springs just for watching them forage for nectar on peony flowers. Of course, we took care as we transferred them from the net into the killing jar. That's how I learned how important the peony's reward must be to the young daughters and larvae back in the expanding nest as hornets and paper wasps don't collect or eat pollen. When I shoved a mounting pin through a wasp's thorax it leaked nectar. I must have punctured the storage crop. The observation suggested that wasps took peony nectar home to share.

Everyone who gardens or picnics in the Northern Hemisphere has wasp stories. As they hunt for food, a wasp burns energy in flight and they need sugar to continue. That's why we often see them taking nectar on flowers with petals forming shallow cups. Social wasps typically have tongues far shorter than honeybees and bumblebees, and there's plenty of sugar in ripe fruit. That explains why it's often unwise to pick up windfall plums or peaches.

However, few social wasps are content with dessert. They want meat to feed larvae remaining in paper combs. Insects in the meadows and shrubbery are their legitimate prey. Most North Americans have long memories of hornets and yellow jackets inspecting the contents of our barbecues, left to cool on paper plates. As we will see in the following chapter, while some of the larger hornet species are merciless raiders of beehives, some may become pollinators of orchids as the season progresses.

When it comes to the evolution of insects, it is correct to say that all bees are wasps but not all wasps are bees. The true families of bees are placed in the suborder Apocrita. This also includes all the social wasps I've described above as well as solitary, parasitoid wasps such as ichneumons and braconids. As the stanza of the poem that began this chapter suggests, wasps and bees share some features but how do they diverge? True wasps are narrower and each has an exaggerated wasp 'waist' linking its thorax to its abdomen. As mentioned above, wasp bodies are programmed to collect and feed meat to other females and offspring. In particular, colonial or social wasps catch and kill their prey, chew it into pastes, store this pablum in their crops then regurgitate it into the mouths of hungry larvae. This process is called trophallaxis and if you find the thought of mouth-to-mouth transfer of meat slurry repugnant remember it's much the same process in most birds including adorable penguins

Bees, in contrast, have a much reduced, unobtrusive waist giving them a rounder, 'pear-shaped' figure. They raise their sisters and offspring on a diet of bee bread consisting of a pollen loaf mixed with nectar and in a few species (described in Chapter 1) it's closer to a pollen and oil mayonnaise. The domesticated honeybee (*Apis mellifera*) goes one-step further in the earliest stages of childcare. The nurses feed the youngest larvae liquids manufactured by glands in their heads so it's trophallaxis all over again. Once the larvae enlarge and develop they are big enough to digest bee bread.

It looks like a perfect subdivision between brutal barbarian wasps and sweetly vegan bees, but it's not that simple. Bees and wasps are too closely related. At least three tropical American bee species in the genus *Trigona* are called vulture bees because they strip animal carcasses and will even forage on dung. Yes, they do produce honey pots but most of the rural and indigenous people think they taste awful for obvious reasons. In contrast, consider the species of single mothers in the pollen wasp family Masaridae. They are found in Africa and the Western Hemisphere. These wasps store pollen and nectar in their crops, often visiting flowers of the snapdragon (Scrophulariaceae) and mint (Lamiaceae) families, as well as scorpion weeds (*Phacelia*), to provision their larvae. Respectable pollen wasps would be revolted by the diets of their distant, vulture bee relatives.

We had to think hard about all the information in the preceding paragraphs as we completed field and lab studies. Our social wasps were known caterpillar hunters. It's a soft and juicy prey, and relatively easy to chew up and turn into paste. In fact, the flower buds of Brown's peony were plagued by at least three different species of caterpillars all belonging to the family of owlet moths (Noctuidae). The mother moth laid an egg on the flower bud. When the caterpillar hatched, it drilled a hole through the tough sepals ultimately devouring the ovaries and other organs within. When the remaining segments of the bud opened, an obese caterpillar was exposed on the mangled remains of the blossom.

I would love to be able to tell you that explains the evolution of the colours, chemistry and floral presentation of this atypical peony. The plant should attract wasps initially to rid itself of herbivores that spoil its chances to reproduce. In exchange, the healthy, surviving flowers reward the predators with an energy drink. Wouldn't that confirm a mutually beneficial relationship? It's a reasonable theory. Most of the prettier species of peony have glands on their leathery sepals while they are in bud. Their secretions feed ants (related to wasps) and ladybird beetles, which keep peony blossoms clean of mealy bugs and other sucking pests. It's not unique to peonies either. Extra-floral nectar glands evolved independently on the leaves and stems of many unrelated plants including elderberry (*Sambucus*), vetches (*Vicia*), cotton (*Gossypium*) and passionflower vines (*Passiflora*). In the next chapter, I have more to say about the importance of nectar glands exposed on the branches of Australian acacias.

There was only one problem with my theory. In the years we watched wasps none of us observed these killers hunting and catching caterpillars on Brown's peony. I'm not saying it never happened. Perhaps it occurred while caterpillars were tiny and big wasps snatched them up in their massive mandibles far too fast for us to follow. Maybe someone else will report wasp hunts on Brown's peony in the future. On the other hand, maybe there's a more subtle explanation and we need to look again at the generous chemistry in the liquid pumped out of those green nectar glands.

## PEONY NECTAR AS A WASP DIETARY SUPPLEMENT

Let me reiterate, you can't make a baby wasp entirely out of sugar and water. You need amino acids that link up to form proteins and enzymes that make the framework of insect cells and push the cycle of development along until a larva is big enough to undergo her own metamorphosis. That's why wasps pursue meaty prey. But how do you nourish infant wasps so early in spring when soft grubs and caterpillars are small and/or uncommon? At least, here in the scablands, Brown's peony is ready to offer wasps similar nutrients without stalking, killing and butchering prey. The sugar in the nectar benefits winged big sister but little sister in her comb cell needs special amino acids. Did you know that wasps, like moths, must spin a cocoon when they are ready to mature into winged adults and the coiled threads of silk are mostly protein? You can munch up a struggling caterpillar if you prefer, but the alanine, serine, lysine and glutamic acid in wasp silk are already dissolved in the peony's nectar. Why hunt for a few caterpillars for available protein when drinks rich in amino acids are free and require no effort?

In any case, the ovaries in a typical Brown's peony flower swelled into dry fruits called follicles following visits from the big hover fly and/or wasps. While an ovary contained about 18 unfertilised seeds, there was never enough room in an ovary for all of them to mature after pollen tubes fertilised them. When each follicle split open along one seam it released, on average, four plump seeds.

While checking one of the classic books on pollination we soon learned that we'd described a natural system that is still regarded as uncommon but not unique. Most of the research in the 19th and 20th centuries on pollination by social wasps was completed in Europe. Favourite flowers of hornets and paper wasps included ivy (*Hedera helix*), some helleborine orchids (*Epipactis*) and one of the odder members of the snapdragon family known as the common figwort (*Scrophularia nodosa*). These three unrelated plants made flowers sharing three features with Brown's peony. First, no naturalist was enthusiastic about the flowers' smell. In fact, there was little mention of odour. Second, all lacked the vivid yellow-blue-purple colour patterns associated with bees. Instead, these blossoms dressed in browns, greens or a mottled combination of both. The older generation of naturalists described the colours of wasp flowers as dull. Finally, and of equal, if not greater importance, these flowers pumped out great droplets of nectar in wide and easily accessible bells and cups or exposed them on lip petals. I have more to say about the Eurasian helleborines in the next chapter.

Perhaps this is why we don't see flowers pollinated primarily by large social wasps more often. Yes, the biochemistry of colour is cheap and easy to make. Yes, these peonies don't appear to spend much of their budget on scent, but how many plants can afford to give up so much water to make such a big nectar reward? Sugars are a simple product of photosynthesis, and their daily concentrations in the nectar of Brown's peony were not extravagant compared to analyses of nectars secreted by flowers pollinated by bees or birds. What we can't answer, yet, is the expense of seasoning nectar with so many amino acids which most animals, from wasps to humans, find essential to their growth and survival.

## CHINESE FELWORTS AND GERIATRIC WASPS

As research continues, it becomes exciting to speculate on how fine-tuned wasp–flower partnerships might be around the world. Maybe big, brutish social wasps are more common pollinators than we thought. Maybe we haven't been looking in the right place or season. As I write this chapter, my Chinese colleague, Dr Zong-Xin Ren, is investigating hornets and flowers in the hills around the city of Kunming, Yunnan. By late September, these insects stop killing bees, their preferred prey, because production of new sisters within their paper nests is coming to an end. Raising more workers is a waste of resources once the new generation of queens mature, mate, fly off and prepare to hibernate until spring. Freezing temperatures will kill off all the inhabitants of the old colony. What remains for the surviving wasps? Speaking from personal experience, sugary treats offer some consolation after you are retired and underemployed.

(Left) Flowers of a Chinese felwort (*Swertia bimaculata*) showing scattered nectar guide spots and two nectar glands on each petal. (Right) An ageing female wasp (*Vespa bicolor*) takes nectar from a flower of a Chinese felwort (*Swertia kouitchensis*). Photos by Zong-Xin Ren.

In Yunnan, the remaining and ageing workers spend their final days drinking nectar provided by autumn flowers, especially felworts (*Swertia*). These bushy herbs are not related at all to peonies and belong in the gentian family, Gentianaceae. Each felwort flower has two shallow nectar pits at the base of each of its five flat petals, easily accessible to the short tongues of thirsty hornets. Perhaps flowering in autumn offers some plants a natural advantage luring a once implacable hunter into a short, and late career move as a pollinator. The felwort pays an ageing hornet in nectar in exchange for pollen transportation instead of leaving her to drown her disappointments in the juices of fallen fruits. In north-western America, Brown's peony partners with hunting wasps as their colonies begin to expand. In south-western China, felworts console them in their twilight. Both end in the production of a new generation of seeds. I like the West versus East symmetries, but wonder if both systems ever exist in different seasons on the same continent?

Returning to GROWISER, it was a wonderful place to work and I recommend it to anyone who wants to see some of the best populations of some of the rarer wildflowers of north-western America. Its population of Brown's peony allowed us to make an addition to an infrequently visited topic in plant reproduction. It never occurred to me I'd stumble across pollination by social wasps again, but I did. That meant a road trip through Shangri-La in Yunnan. Yes, the county of Shangri-La does exist, but no one there sings those old Burt Bacharach songs. I'll explain all that in the next chapter giving us the opportunity to bring back orchids, allowing us to jump from China back to Australia.

# 7

# A corsage for wasps

*Petruchio: Who knows not where a wasp does wear his sting? In his tail.*
*Katherine: In his tongue.*
*Petruchio: Whose tongue?*
*Katherine: Yours, if you talk of tales, and so farewell.*
*Petruchio: What, with my tongue in your tail?*

– The Taming of the Shrew, *Shakespeare*

THE CHINESE STATE OF YUNNAN IS A CENTRE OF CULTURAL DIVERSITY AND some of the world's most beautiful wildflowers. At least 24 minority groups share the Hengduan Mountains straddling the state. The Nakhi or Naxi people are most numerous in north-western Yunnan. Their religious beliefs are influenced by the Tibetan branch of Buddhism, but many continue to practice the rituals and teachings of the Dongba, a wise man from eastern Tibet known as Dongba Shilo. Legend has it he lived in a mountain cave ~900 years ago. The Nakhi people continue to venerate a slope where many believe he lived. It's one of their major shrines and is set above the village of Baidi. People come to pray and make fires at the base of surrounding pine trees, to perform a sacrifice, and have a picnic. The pines do not seem to suffer from the fires and are surprisingly tall.

On 6 June 2015, our botanical team decided to visit the shrine. We needed lunch and a break as our field sites were hours away in the county of Shangri-La. To enter, we had to buy tickets and walk up the hill. The first part of the trail took us past tethered Nakhi ponies wearing saddles or embroidered blankets. These people are known for their needlework and horsemanship, but this exhibition was merely a photograph and pony-ride opportunity for tourists. The trail of concrete and stone bordered an arbour of native vegetation of oaks, yellow-flowering dogwoods (*Cornus*) and ferns.

The nice stone path gave way to a stairway of rotting logs. What I did not expect was that the shrine was actually a series of natural and interconnecting mineral springs. As water rushed downhill, it resembled milk, leaving layers of white, thick deposits in the natural spillways and gutters. Pools on the plateau were clear to the bottom and supported tadpole nurseries and green threads of a *Spirogyra*-type algae. The water was surrounded by shrubs of short scrubby willows (*Salix*) and barberries (*Berberis*) mixed with sedges and reeds.

It was here that my Yunnan colleague, Dr Zong-Xin Ren, pointed to more than 200 flowering stalks of helleborine orchids (*Epipactis mairei*). Each stalk bore six to eight flowers held horizontally or nodding. The flowers were pinkish-yellow, but the thickened sculpture or callus found on each lip petal was dark burgundy or maroon, rather reminiscent of the petal

A stalk of the Chinese helleborine *Epipactis mairei* at the shrine in Shangri-La. Photo by Zong-Xin Ren.

colours of Brown's peony in the preceding chapter. Yes, nectar was oozing out of the base of the lip and that puddle lay under the rod of the fused male and female organs, known as a column.

Unlike the mosquito-pollinated orchids in Chapter 3, helleborines do not hide their sweet drinks in hollow pouches, tubes or spurs. In a few species of hellborines, the lip petal forms a shallow cup for nectar but, in any case, droplets ooze from the base of the lip and are easily consumed by insects that lack long tongues. I could see these driblets without the benefit of a magnifying lens.

Ren and his then assistant, Zhi-Bing Tao (now Dr Zhi-Bing), went off to interview some of the Nakhi pilgrims at their fires under the pines. I stayed behind to photograph the orchids with my mobile phone, I was in for more drama. A wasp was crawling about a flower stalk, taking nectar from each flower. When I tried to take its portrait, it rose slowly into the air. I yelled for Ren (still within earshot). He caught the wasp in a little organza mesh or bonboniere bag we use to isolate flowers we want to use later for breeding experiments. We placed the bag in a killing jar. Ten minutes later Mr Tao bagged a small fly with shiny black and yellow bands. This was a hover fly (Syrphidae) and another bee mimic, a cousin of the much larger and hairier *Criorhina* I caught on the peony in the preceding chapter. Why were the wasp and hover fly moving so slowly that my colleagues could nab them in such little bags? My original guess was that both were burdened with heavy deposits of the orchid's pollen. Like the midge and blunt-leaved orchids in the earlier chapters, helleborines release their pollen grains massed together in discrete lumps known as pollinia. Bearing a heavy load of plant sperm is the price of a healthy drink. I would learn later, that was not the whole or the most honest story.

We took photos of the dead wasp. Its body was coloured boldly in black with red bands and splotches. Even the wings were sooty with highlights of salmon pink. The cells made by the veins in the wings were visible and that's an important tool in insect identification. I sent photos to Dr James Carpenter, a wasp expert at the American Museum of Natural History in New York City. He confirmed it was a paper wasp like some of the ones I'd caught on Brown's peony. Dr Carpenter was reasonably certain it was *Polistes tenebricosus.* Snap! That's another link between a homely American peony and a Chinese orchid.

Ren pinned and labelled the wasp and it remains in a walk-in freezer room with all the pollinators he's collected on the wildflowers of Yunnan over the years. I picked a few orchid stems to press as museum vouchers as we would need to provide evidence of both the plant and the insect. Once dried, they were given to another colleague, Professor Yi-Bo Luo, in Beijing. While we will see that the pollination of hellborines has been under investigation since the 19th century, the *P. tenebricosus* carrying the pollinia at the Nakhi shrine was a new record for *Epipactis* orchids. When we left the shrine we headed for a nearby restaurant and celebrated our catches over a traditional chicken hotpot. I asked Ren what did the Nakhi pray for under the pines? He replied, 'The men prayed for rain and the women said they prayed their children would do well on university entrance exams'. Somethings will never change regardless of country, culture or language. Farmers want more rain, their wives don't want their kids to be farmers, and unwanted scientists pursue specimens in unexpected and often inappropriate places.

I might also add that wasps will always pursue orchids but you have to be at the right place at the right time. While our collection of a Chinese paper wasp on a Chinese helleborine was new to the story, these insects and flowers go back to Charles Darwin as he figured out how English helleborines released and received pollinia in the 19th century. What Darwin never learned was that some helleborines offer more than lots of easily quaffed nectar.

## HELLEBORINES AND THEIR WASPS

With ~70 described species in the Northern Hemisphere, few orchids appear to be as self-obsessed with turning every flower into a seed-filled capsule as the helleborines. While many orchid species produce long-lived flowers that stay in bloom for weeks until exactly the right pollinator comes along, helleborines aren't one of them. They are far less picky about when and where they will have sex. After invading urban areas, devoid of typical pollinators, the broad-leaved helleborine (*Epipactis helleborine*) of Europe evolved repeatedly into self-pollinating races or subspecies. Otherwise, native helleborines employ a variety of colour and scent tricks to entice pollinators. Observations of wasps, syrphid flies or a combination of both are common enough, but field studies in Poland, the Czech Republic and Lithuania report that flies representing other families, bumblebees and even a few beetles may patronise helleborines and carry their pollinia. In Thailand, climbing crickets (*Homoeoxipha lycoides*) are among the most frequent visitors and pollinia transporters of a yellow-flowered helleborine (*E. flava*).

While there's always an easy nectar reward for insects in a helleborine flower, it may not be the only reason some visitors land on the lip petal. Some helleborine blossoms offer gross and deceptive advertising, which fools young mothers and ends in infant mortalities. Wasps are not the victims, though. Female hover flies (Syrphidae), like the one I caught at the shrine, take the orchid's nectar, as usual, but then they are also deceived by the sculptures on the flowers as produced by some helleborine species in China and other parts of the world. While most adult hover flies live on a diet of pollen and nectar, their offspring eat succulent flesh. Pregnant hover flies lay their eggs in colonies of aphids that suck juices out of green plants. Once the hover fly maggot hatches, it becomes a ravenous wolf among the fold.

Consider the Middle Eastern helleborine orchid, *E. consimilis.* In Israel, these wildflowers offer the usual drinks to female hover flies, but the flowers bear dark warts. Male flies land, drink and carry off the pollinaria, while females show an extra behaviour that makes them far more dependable as pollinators. These female flies drink but also lay their eggs among the warts mistaking them for colonies of dark-coloured aphids. While female flies also pollinate the orchids, their offspring starve to death upon hatching on the flower. Fake aphids are not digestible. Some ecologists think the helleborine orchids produce smells like aphid colonies similar to the Chinese slipper orchid (*Cypripedium subtropicum*) seen in Chapter 2.

This is definitely the case in a second helleborine orchid flowering through eastern Asia and in our study sites in the Hengduan Mountains. *Epipactis veratrifolia* pulls a similar trick but reaches a different level of specialisation. Their flowers have the usual warts but they also become infested with real, live aphid colonies. These orchids may in fact attract winged and pregnant aphids to their buds as their flowers smell like more aphids. Both the orchid flower and the real aphids produce the scent molecules of pinenes, limonene and eucalyptol. In turn, this double concentration of aphid and orchid perfumes attracts female hover flies. The Chinese *E. veratrifolia* attracts pregnant pollinators that also help rid the plants of sucking pests when they lay eggs in the flowers. In this case, there is enough real prey for the carnivorous maggots to gorge on long enough to metamorphose into a new generation of hover flies.

Likewise, a helleborine may offer nectar to a predator but needn't smell precisely like its preferred prey. Instead, the flower may secrete a scent like leaves damaged by caterpillars. This is the case in wilder, country forms of the European *E. helleborine*. The flowers give off the smell of green leaf volatiles that are usually released after caterpillars bite and chew the foliage of cabbages and other plants. Consequently, the European helleborine is most likely to be pollinated by common species of European yellow jackets instead of pregnant hover flies.

Now let's review that sluggish behaviour of my wasp feeding on helleborine flowers by the shrine pools. Appearances may be deceiving. We should first consider an earlier and famous study conducted by two Danish ecologists, Bodil K Ehlers and Jens M Olesen, of wasps returning to the violet helleborine (*E. purpurata*) of Western Europe and the broad-leaved helleborine (*E. helleborine*) of Eurasia and North America published in 1997. The nectar of both European helleborines intoxicated their wasps due to their pollinators' unhygienic dining habits. Social wasps love overripe fruit and rotting windfalls that are colonised by microbes including a blackish mould (*Cladosporium*). Sloppy, gluttonous wasps carry the mould's spores on the fruit's skin to the nectar in the helleborine flower. The nectar of the same flower may also been contaminated with additional airborne yeasts (*Candida*). As the moulds and the yeasts grow in the nectar they digest its sugars producing alcohol.

There are those who would argue that the orchid, yeast and mould form a short-term partnership and enjoy mutual benefits. The nectar gives fungal spores a place to develop and reproduce. The same nectar attracts a wasp that will spread yeast and mould to new flowers or overripe fruits. In return, yeasts and moulds change the contents of the nectar so much its consumption dumbs down the pollinator. A slow wasp spends more time in the same patch of orchids and may be more likely to collide with male and female parts on a flower's column as its foraging speed slows.

Of course, we can't say whether the wasps are enjoying a guilty pleasure. It's a little like one of the prose poems by the humourist Don Marquis (1878–1937). He wrote about a dissipated hornet addicted to eating flies made drunk after they sponged up booze spilled in illicit saloons. The hornet, who calls himself Crusty Bill, says ...

> *the curse of drink got me the mad life began*
> *to tell upon me I got so I would not eat a*
> *fly that was not full of some strong and heady*
> *liquor ...*

How would my Australian friends and in-laws have reacted to the sight of my paper wasp struggling bravely to stay airborne under an unfair load of pollinia? I suspect they would have said she was 'three sheets to the wind'. The relationship between native wasps and orchids in Australia has its own deviant streak, though, and I promise to discuss it in depth later.

My observation of the wasp and the Chinese helleborine was not unique. There are probably many more orchid species in China luring social wasps into transporting their pollinia. Some systems are simple, while others show a remarkable chemical sophistication. At least two operate on pure deceit exploiting the hunger of their pollinators.

## CHINESE ORCHIDS FOOL WASPS

The fringed coelogyne (*Coelogyne fimbiata*) grows on shady rocks or tree trunks from India through much of south-eastern Asia, preferring subtropical–tropical latitudes. Aside from the pretty, fringed margins of its lip petals, there's nothing spectacular about its greenish-yellow flowers except that the white lip petal bears a contrasting, red wine splotch. The same flowers smell strongly of ripe pears and freshly cut grass. Its late flowering period from the end of August into mid-autumn is atypical for most temperate zone orchids. It should be noted, though, that this is the time of year in which wasp nests need quite a lot of food to support the rearing of new queens and the males that will inseminate them.

The only visitors to the flowers of the fringed coelogyne in the Yachang Orchid Nature Reserve in subtropical regions of the southern state of Guangxi, China, belong to a species of *Vespula*. Many despise these insects and call them yellow jackets. If you are allergic to their stings a summer picnic or orchard harvest in autumn can turn deadly. Yellow jackets consume sugary and protein-rich foods without our invitation. The fringed coelogyne invites them with its delectable (for a wasp) scent, but the yellow jacket is deceived. The flowers contain no nectar. Yet the wasps cross-pollinate the orchids if they visit more than one empty flower on more than one plant. A team of scientists observed wasps visiting fringed coelogyne flowers in Yachang Reserve 29 times. In all but one visit, each yellow jacket either left a flower carrying a pollinarium on her back and/or deposited pollinia she was already carrying following previous visits to other flowers.

This food mimicry is so common in orchids in general that the fringed coelogyne is merely following a version of the same trend documented repeatedly in other orchids pollinated by deceived bees or flies (Chapter 2). Yellow jackets may be more brutal as hunters or scavengers, but they are no less gullible. Experiments using the typical Y tube used to determine choice in animal behaviour experiments showed that the wasp didn't have to see the flower to find it. It crawled into whichever arm in the Y tube contained the flower's scent.

Smelling like ripe fruit to lure foraging wasps is rather unsophisticated compared with another orchid, the Chinese dendrobium (*Dendrobium sinense*). Although confined to well-populated Hainan Island, off the coast of southern China, the orchid wasn't described as a new species until 1974 as it lives on branches hidden away in mountain forests. The flowers are pleasant by human standards. They are waxy white in colour with a golden-yellow or orange-red splotch on their lip petal. It's charming but there are at least 1,800 *Dendrobium* species in the Asian tropics, New Guinea and Australia with far more spectacular blooms. Many have been hybridised and commercial florists sell them as spray orchids. The flowers of some dendrobium hybrids are so popular they are plucked from their stalks and strung into Hawaiian leis or are presented as an inedible garnish on fruit salads at buffets.

Why should a team of scientists focus on one orchid species restricted to a small Chinese island? The flowers are visited infrequently by bees and butterflies, but they are never the pollinators. This orchid is pollinated exclusively by foraging females of black shield hornets (*Vespa bicolor*). Yes, these insects stop at flowers to drink nectar, but they are really ruthless predators killing a variety of insects to feed to their larval sisters. More importantly, they will gang up on hives of the useful Asian honeybee (*Apis cerana)* to murder workers.

I say useful because until recently the Asian honeybee was the traditional source of honey and beeswax for the Chinese, but it was never domesticated. In the spring, villagers went into the woods, found swarms on tree branches, took them home, and introduced them to artificial hives. Aside from their unusually tasty honey, farmers depended on them to pollinate orchard fruit, garden vegetables and even some medicinal herbs. Guarding these bees from hornet attacks was a priority. My research colleague Dr Zong-Xin Ren grew up in a farming community in the Hengduan Mountains. He remembers summer vacations in which his father gave him firecrackers to scare the hornets away from the hives.

What happens if you are a worker honeybee and there isn't a kid near your hive lobbing explosives at your enemy? You have to depend on your own sisters, of course. You release an alarm smell into the air to warn them. Hive bees, whether they are European or Chinese, warn each other with the molecule (Z)-11-Eicosen-1-ol, a long-chain fatty alcohol. The good news is that secreting this warning helps to mobilise your sisters. The bad news is that it allows the hornet to recognise you as vulnerable and edible.

Just as yellow jackets find flowers of the fringed coelogyne by its scent, so do black shield hornets when the Chinese dendrobium blooms. This time though, the orchid is offering the false promise of a juicy bee instead of a juicy pear. Once again, if you give the hornet the opportunity to choose between the two branches in a Y tube it will take the path towards the branch containing sufficient concentrations of the warning scent. Jennifer Brodmann observed the orchids in bloom in the Hainan Bawangling National Nature Reserve for over 120 h and watched the hornets visit the flowers 35 times. The predators didn't land gently on the lip then crawl into the flower to suck nectar, which it lacks. They pounced on the golden-reddish spot as if it was prey, but left after one second. They'd been fooled long enough for the orchid to dump its pollinia on their backs as they backed hastily out of the flower. Back at the hornet's nest, the research team counted 277 worker wasps entering and 30 were still carrying the orchid's pollen lumps.

Smelling like a distressed insect has served orchid evolution more than once and is probably far more common than we've anticipated. If the pollinator lives on a diet of meat and the sugar in flower nectar at different times over its lifespan, natural selection may push scent chemistry in different directions. Sometimes the flower offers something to drink or nibble attracting the pollinator with a more carnal odour. Consider again the case of some of the helleborines and especially the slipper orchid (*Cypripedium subtropicum*) in Chapter 2. The outer surface of the flower's shoe-like lip petal offers edible hairs to hover flies. They also smell like terrified aphids, the exact place where a hover fly female wants to lay her eggs.

The fringed coelogyne and the Chinese dendrobium are typical extremes on the other side of the bell curve. Their flowers don't make nectar or edible hairs, only false promises exploiting the aromas of sweet fruits and scared bees, respectively. As the Chinese learn more about their unique biota these orchids may eventually come to symbolise the best time of year for a seasonal and local treat. Remember, these orchids bloom as the production of offspring increases in the larger wasp nests. Those paper nurseries are collected and harvested in some parts of Yunnan and other provinces. I don't know how they get rid of the stinging adults, but the nests are sliced into rounds and sold in the stalls of open-air markets in cities, such as Kunming. The larvae and cocoons are still alive. I've visited those markets and watched the exposed and immature insects pulsing synchronously like fans at a football game doing the wave. I understand they are taken home and fried. Perhaps this compensates some apiarists for the loss of their bees.

## ENTER THE GENTLEMEN OF SPRING

The natural history of wasps and flowers described here may one day link to some of the noblest arts in China. For millennia, the Chinese have used four plants as ambassadors of the four seasons. Writers once compared the four species to Confucian gentlemen of refined taste and wisdom. They are depicted on scrolls, screens, fans, inked prints and in poems through to the present day. Bamboo should make us think of the luxuriant and productive summer. Autumn is the chrysanthemum and winter is a native plum (*Prunus mume*). An orchid heralds the spring, but not all orchids are equal.

The soubai or song mei (*Cymbidium goeringii*), and maybe one or two of its sister species, are the favoured gentlemen of spring. They are prized because they are cold-tolerant and can

A Chinese soubai or song mei orchid in a traditional pot. Photo by Weichang Huang.

be kept for decades in traditional clay pots on pedestals outside in the family courtyard. Their leaves are long with generous and sinuous curves serving as graceful models for generations of artists. When it comes to their flowers, though, a Westerner, like myself, finds them nice but anticlimactic. Compared to so many orchid species native to China with their dramatic and voluptuous flowers the blossoms of a soubai seem modest. They are small and their natural colour patterns vary in shades of yellowish-green with white, while some display purplish-brown veins on their lip petals.

What is the attraction that supports such a lucrative but specialised horticulture? I was a guest at a national Chinese conference on soubai in Beijing years ago attended by growers from all over the nation. My hosts walked me down rows of display tables offering every hybrid, mutant and colour variation based on the same species. The containers used by the owners seemed more variable, at times, than the orchids. When I examined the flowers, I employed a hand lens in the hope of catching some subtlety.

I would have been better off using my nose. For the Chinese, the pleasure of these flowers lie in their scent. The great scent biochemist Roman Kaiser attended an orchid show in Tokyo in 1991 taking the time to 'trap' the odours of a soubai in bloom. His analysis identified 30 chemical constituents. The dominant molecules were (E)-Nerolidol at 58% and (E, E)-Farnesol at 11.5%. Both belong to respectable families of essential oils produced by several plants and some fungi. They are also made synthetically by flavour and cosmetic industries for many commercial products. The scent of nerolidol reminds us of fresh tree bark, while farnesol is identified for its 'volatile green' or leafy notes.

This is culturally important. The Chinese often say they like the smell of soubai flowers as they are reminiscent of a freshly mown lawn. Didn't we find that the flowers of the fringed coelgyne had a grassy note? Yes, and a few Chinese botanists suspect that the soubai is also pollinated by wasps. Perhaps a venerable tradition in art and horticulture hangs on scent cues one flower sends to social wasps.

## THE AUSTRALIAN VERSION OF WASP POLLINATION

To provide a contrast, now is the best moment to extend and flip the wasp story to some of the 1,300 orchid species native to Australia. It's a place where you will see wasp-pollinated orchids in bloom consecutively from late winter to mid-summer. Orchid flowers in Australia were not driven by the selective hunger of social hornets or paper wasps. While much of the natural history of Australia's wasps also revolve around females hunting big game for their own offspring, they are largely solitary insects and don't provide meat for other females. However, by my crude calculations, solitary wasps, representing at least three different families, may be pollinating a quarter to a third of all orchid species found in the temperate southern half of continental Australia and Tasmania. Something new must be added here. Almost all the wasps pollinating Australia's orchids are males.

Once again, it's a matter of flower nectar. In this case, Australian plants might be offering too much. In *Where Song Began*, a ground-breaking book by biologist Tim Low, he reviews the evolution, classification and ecology of birds distributed throughout Australasia. Low explains that woody Australian plants have sugar to spare because they grow in soils so low in important mineral supplements that they can't use all the sugar energy they accumulate to build and run other processes performed by plants in other parts of the world. A lot of the excess sugars made by Aussie trees and shrubs are diluted with water and pumped into their flowers to feed pollinators. It is consumed easily, because it is offered in the most shallow floral cups and brushes. We see such flowers in the families Myrtaceae (honey myrtles, eucalypts, tea-trees etc.), Proteaceae (grevilleas, snotty gobbles, banksias, telopeas) and Asphodelaceae (grasstrees) dominating southern woodlands and forests. A surprising number of native species of insects, birds, bats (Chapter 10) and even mouse-sized marsupials live primarily on a diet of nectar yearlong. Different plants offer drinks in all four seasons, as winters are mild in the south, so there's always something in bloom.

Australia's wasps then have plenty of easy options. In fact, they don't even need to seek shallow flowers at all to drink cocktails of sugars, amino acids and triglycerides, if so inclined. Australia's wasps can ignore nectar-filled flowers by visiting hundreds of species of wattles (*Acacia*) because these woody plants have found yet another use for those excess sugars. Ironically, the puffy yellow flower balls and fuzzy sausages on wattle bushes never secrete nectar. Bees, flies and beetles visit them only for pollen. Instead, the same trees and shrubs pay for bodyguards with nectar glands on their foliage, much like the Eurasian peonies in the previous chapter. Ants and wasps protect the wattles as they develop new, soft vegetation or tender flower stalks through winter and early spring. In the mid-1980s I had a look at nine wattle species. Four of them secreted leaf or stem nectar and they fed 58 wasps I caught representing seven families of solitary insects.

The most common wasps I netted belonged to the flower wasp family Thynnidae, and I learned a valuable piece of natural history. The sugar secretions of the foliage of golden wattle (*Acacia pycnantha*) were especially popular with these wasps and several of them arrived on wattle stems at the conclusion of their honeymoons. Males and female thynnids fed while attached to each other. The grooms are long and winged. The females are short, plump and wingless. She spends most of her life underground. When a female flower wasp is ready to mate, she crawls up on a grass stalk or fallen twig and broadcasts her sex pheromones into the air. An incoming male attaches her abdomen to his genitalia and off they go on a nuptial flight ending with a nutritious drink in a flower cup or on a wattle leaf gland. The winged groom drinks first then usually feeds his flightless bride nectar by the same process of trophallaxis we saw female social wasps and hornets use on larvae in the previous chapter.

If you are an Australian thynnid wasp though, there's no devotion to the nuclear family. The male will dump the female after she's had her fill. Upon returning to the soil, she now has enough energy and nutrients to give her the strength to find beetle grubs, sting them and lay her eggs in them. Ultimately, the lifecycles of most of the wasps pollinating Aussie orchids depend on females parasitising and paralysing host larvae. Grinding prey into easily digestible meat pastes isn't an important maternal skill here.

Some leek orchids (*Prasophyllum*) in Australia follow the annual mating cycles of the flower wasps I caught on the wattles. Males with or without brides are often found on these

A male wasp inspects the lip petal of a carousel spider orchid (*Caladenia arenicola*) in Western Australia as he is still copulating with a wingless female of his species! Photo by Mark Brundrett.

wildflowers that are so tall, and have such asparagus spear thick stalks they are known as king leek orchids. In all leek orchids, the flower buds don't twist upside down as in the previously discussed midge orchids, so when the petals open, the lip petal sits on top of the flower like a bonnet. Nectar is secreted at the base of a fleshy, green-brownish triangle on the lip, known as the callus. This time, the wasp must stand on his head to drink nectar from the callus and the pollinarium will be deposited on his back. By now, you will have noticed that the whole system is quite similar to the midge orchids in Chapter 4, except leek orchids entice much larger pollinators.

Why should any wasp visit an orchid, though, when so many unrelated plants produce masses of flowers at the same time and the woodland is just one big nectar bar? In fact, most Australian orchids dependent on wasps either downplay nectar or don't offer any at all. Natural selection probably favours deceit when a simple wildflower can't compete with massive flowering trees and bushes dripping with sweets. What all these orchids have in common is that they specialise in manipulating male wasps. In the past two centuries, children and naturalists alike have observed these flowers and given wasp-pollinated orchids fanciful names based on their shapes, sculptures and colour patterns. Australia has its tongue orchids (*Cyrptostylis*), undertaker orchids (*Lyperanthus*), old man orchids (*Calochilus*), hammer orchids (*Drakaea*), bird orchids (*Chiloglottis*), elbow orchids (*Thynninorchis*) and spider orchids (*Caladenia*).

## EDITH COLEMAN AND HER TONGUE ORCHIDS

Interpreting the story of why male wasps are attracted to certain orchids began back in the suburbs of the state of Victoria in the 1920s. Remember, this was a time when there weren't enough professional biologists to deal with the extraordinary biodiversity on the continent. Natural history clubs appeared all over the country and some had their own publications. Some members became adept at describing the behaviours and lifecycles of insects. Today, sex mimicry in Australian orchids is a topic for scientists with sophisticated laboratory equipment and eager graduate students, but all derive ultimately from the work of one woman, Edith Coleman (1874–1951). She made her discoveries in bushland remnants found around towns, including Blackburn and Healesville in Victoria.

Her papers attracted international attention from the scientific community. Mrs Coleman first tackled the red and maroon dappled tongue orchids (*Cryptostylis*). The flowers offered no nectar but the visiting wasps, all males, were so excited they left deposits of their sperm on the lip petals. In fact, the insects were so obsessed with the orchid that Coleman learned she could sneak up on them with a pair of scissors, snip off the flower stalk and both blossom and wasp tumbled into the vial of formalin she held beneath them. How better to preserve the evidence of the pollinaria glued to the back of the wasp's thorax or, in some cases, the tip of the abdomen near the male's genitals. Although Coleman examined the pollination of several species of tongue orchids, all the male wasps carrying pollinia belonged to the same species – *Lissopimpla excelsa* – an otherwise respectable member of the family Ichneumonidae. *Lissopimpla* species are largely restricted to Australia and New Zealand. The females show no interest in tongue orchids. They are winged parasitoids, stinging and laying their eggs in moth caterpillars.

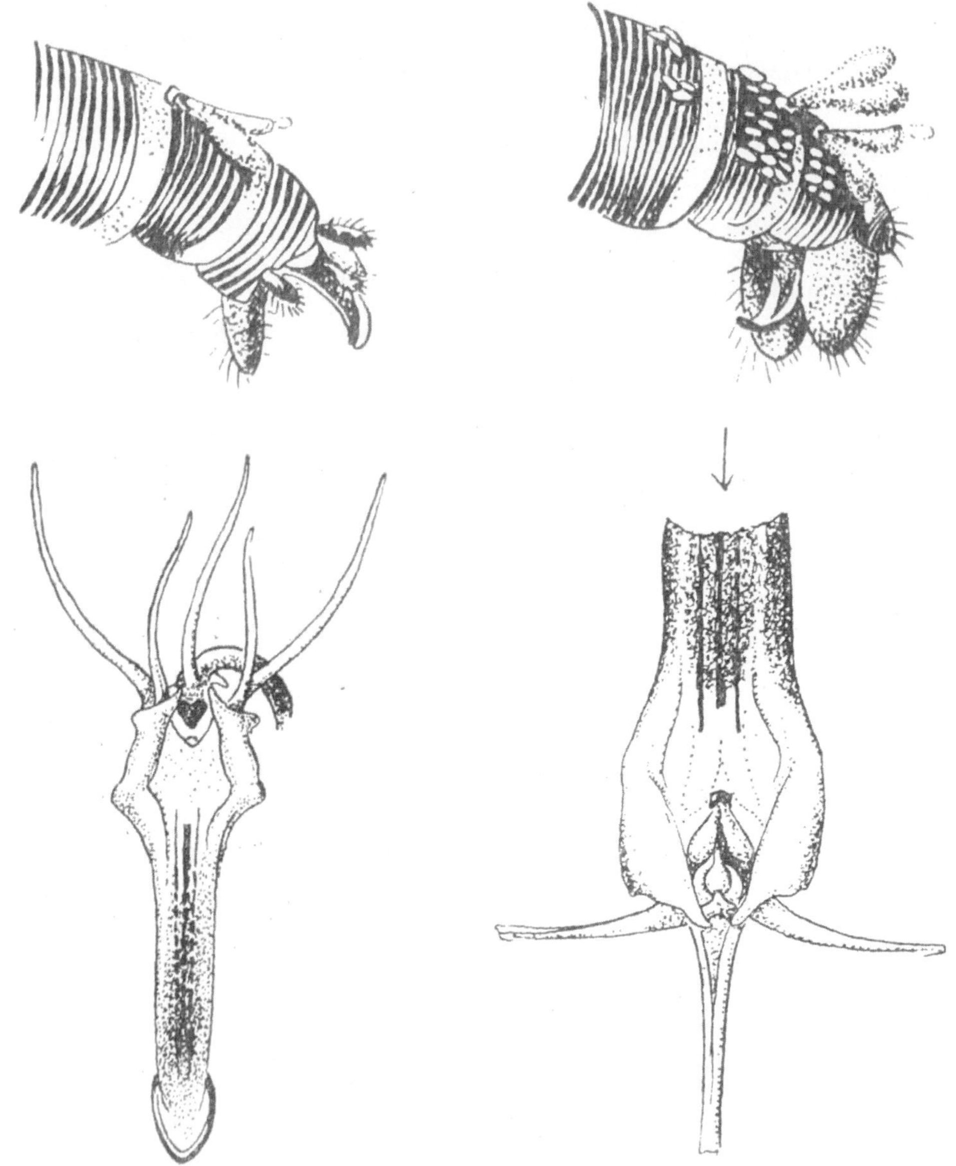

Illustrations by Tarlton Rayment depicting the pollination of tongue orchids (*Cryptostylis*). The top two drawings show the orchid's pollinia deposited on the wasp's abdomen and also shows his genitalia. The bottom drawings show the actual flower. Reproduced with the permission of the Field Naturalists Club of Victoria.

Referring to tongue orchids as an ichneumon wasp's mistress is not a great exaggeration and the repeated response of the male wasp to the flower's charms suggest the orchid is anything but coy. Consider a paper in one of the journals of the Linnaean Society published in 2009 long after Edith Coleman's contributions to science. Marked male wasps proved to be well travelled during their short lives flying over 600 m in search of mates real, or floral. They were seduced repeatedly by western tongue orchids (*Cryptostylis ovata*) blooming in isolated clumps. A lot of cross-pollination occurred. The wasps spread the orchids' genes over the countryside, but these gentlemen were faithful, in their fashion. Based on mapping the orchid plots and marking the males, the investigators learned that half of all the wasps returned to the same orchid in the same patch within a week. For those of you who wish to watch the process, but lack the time and funds to find the best piece of Australian bushland, videos of this process can be viewed on YouTube.

## DIFFERENT ORCHIDS FOR DIFFERENT MALE WASPS

Seduction is specific in the wasp-pollinated flowers of temperate Australia. The patterns on the lip petal of a tongue orchid mean nothing to male wasps in the genus *Campsomeris* in the family of scarab hawks (Scoliidae). Their females prefer to sting and lay their eggs in the grubs of dirt-dwelling scarab beetles and their allies. Male scarab hawks are besotted by flowers of *Calochilus* species, an orchid genus field naturalists find adorable. Known as beardies, bearded orchids, old man orchids and even bearded bolshies (after the caricature of a bomb-toting Bolshevik) the deception of the male wasp takes a new twist. Unlike the sleek females of *Lissopimpla, Campsomeris* brides tend to have thicker, hairier, and more colourful abdomens. Therefore, with a few important exceptions, most of the 28 species of bearded orchids bear a lip petal with a luxurious pelt of rusty-red, rum raisin, green or purplish hair-like structures. Two more sets of structures mimic the body of the female wasp. In some species, there is also a smooth, shiny section on the lip petal located under the flower's column. While botanists call this the speculum (mirror) it appears to represent the thorax of the female wasp where she folds her polished wings. The deception helps to orient the male wasp in the right direction so that the pollinarium in the column's anther is deposited on his face. One thing more is needed to complete the masquerade and make certain that pollinarium transfer is precise. The base of the column stalk has a raised ridge or collar. On the end of each side of the ridge is a rounded, often lustrous structure that looks like an eye. The male wasp looks at and probably sees a potential mate with two pretty peepers. In contrast, the human imagination sees a hairy face in the flower and he's up to no good.

The most recently published account of a male scarab hawk pollinating a flower of the copper beard orchid (*Calochilus campestris*) was published in 1993 by Peter Branwhite and Professor Colin Bower. The male *Campsomeris tasmaniensis* landed on the lip petal, but he wasn't bearing pollinaria at the time. The wasp opened the genital claspers on his curled up abdomen, stopped fluttering his wings and placed his head under the column in between the fake orchid eyes on the collar. As the insect backed out of the flower, the pollinia broke up and fragments fell on his face but lover boy wasn't finished with his flower yet. After a brief rest, he entered the same orchid a second time. This time he left pieces of pollinia on the receptive stigma. Too bad for the orchid if the passionate wasp demands a second coupling. It will only result in self-pollination, not cross-pollination. Observations on the wiry beard orchid (*Calochilus caeruleus*) in 1974 showed that wasps found the flowers even when they were hidden by tall and dense stalks of grass, suggesting that a sexy pheromone, which human noses can't detect, brings in the males.

This balance between the colours and shapes of dummy females found on lip petals and the flower's smell is of major importance to the hundreds of remaining wasp-pollinated orchids in temperate Australia. Most depend exclusively on those flower wasps (Thynnidae), described above, in which females are wingless. Some of these systems are so specialised that only one or two wasp species are known to pollinate one orchid species. How did this become so specific? We are not sure, but a lot of the answer revolves around the sculpture and colour of the dummy female on the lip petals and scent made by the whole flower. The male only recognises the fake female at close range. By human standards, it surprises us they are so deceived. Naturalists and botanists looked at the weird sculptures on the lip petal for close to 150 years and never presumed it mimicked a virginal mate of any insect. In fact, by the mid-20th century some botanists proposed that wasp visitors to spider caladenias were females attempting to sting a prey decoy resembling a real spider. Never ask a botanist to identify the sex of a visiting insect. Male wasps, of course, have no stingers.

It's the long-range smell of the flower that mimics the appropriate female thynnid when she perches on a twig and releases her pheromones. Bird orchids (*Chiloglottis*) were found to produce a special class of scents called chiloglottones. Once again, our noses don't react to these unique perfumes which are based on condensing and degrading fatty acids. The ability of the orchid to produce chiloglottones on special days depends on the flower's daily exposure to the ultraviolet index that is invisible to human eyes. Sometimes two different bird orchid species flower at the same time and manufacture the same chiloglottone molecule. Nevertheless, each species of bird orchid will be visited by a different wasp species because the dummy females on the lip petal of each flower differ in size, shape and colour. In other bird orchids, the scent of the flower probably appeals to only one wasp species because it mixes more than one chiloglottone at different concentrations.

It's also common for some of these plants like the hammer (*Drakaea*) and elbow (*Thynninorchis*) orchids to have a thin joint that attaches the lip petal's stalk to the base of the pollen-making column. This flexible hinge allows the lip petal to change its angle and move up and down when yanked. This occurs when the male flower wasp pounces on the dummy female and decides it makes the perfect mate. He tries to carry the dummy away in the expected nuptial flight. Instead, he crashes into the column and has the pollinarium deposited between his wings on his thorax. Cross-pollination occurs when the same wasp tries to perpetrate the same abduction on the lip petal of a second flower on a second orchid.

## THE ELEGANCE OF CALADENIAS

It's time to consider aesthetics. Few of us would gather a bouquet of Eurasian helleborines or any of the Aussie orchids I've mentioned thus far and present them to a loved one or display them in a fancy vase at a formal dinner party. All of these flowers, as I've mentioned, are small, their colours aren't vivid, and some sort of magnifying lens is required to appreciate them. There's one large exception to the rule represented by more than 230 species known variously as spider, dragon, wispy, clown or green comb orchids, and all placed in the genus *Caladenia*. Even the snobbiest breeders of pampered greenhouse orchids would have to consider their magnificence. Some people look at them and suggest they are the illegitimate offspring of an antique chair and a jewelled brooch. They are protected in Australia but there are always some who can't resist their unique forms and colours, and continue to pick them. Such impulsive greed is sad as the flowers droop rather quickly once separated from their stalks.

*Caladenia* means beautiful glands and refers to the pigmented and ornamented pegs or 'calli' in the centre of each lip petal. Some calli probably release pheromones. People who admire calli may think they look like miniature, upside down golf clubs. There may be only one big, fat lobed peg on the lip or more than a dozen depending on the species. Calli are almost always distinct in colour

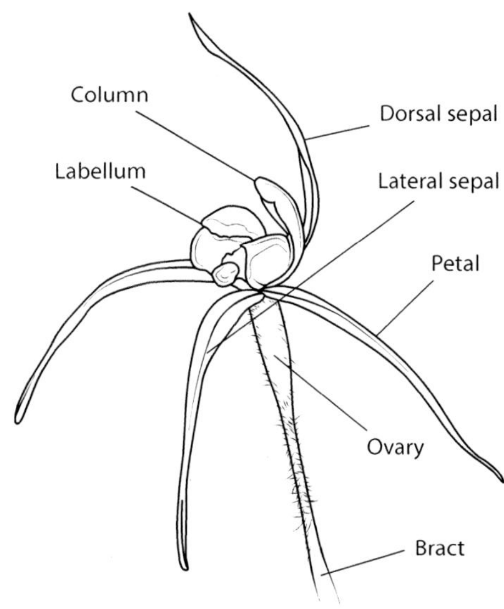


Diagram of a spider *Caladenia* flower. Illustration by Envisage Information Technology.

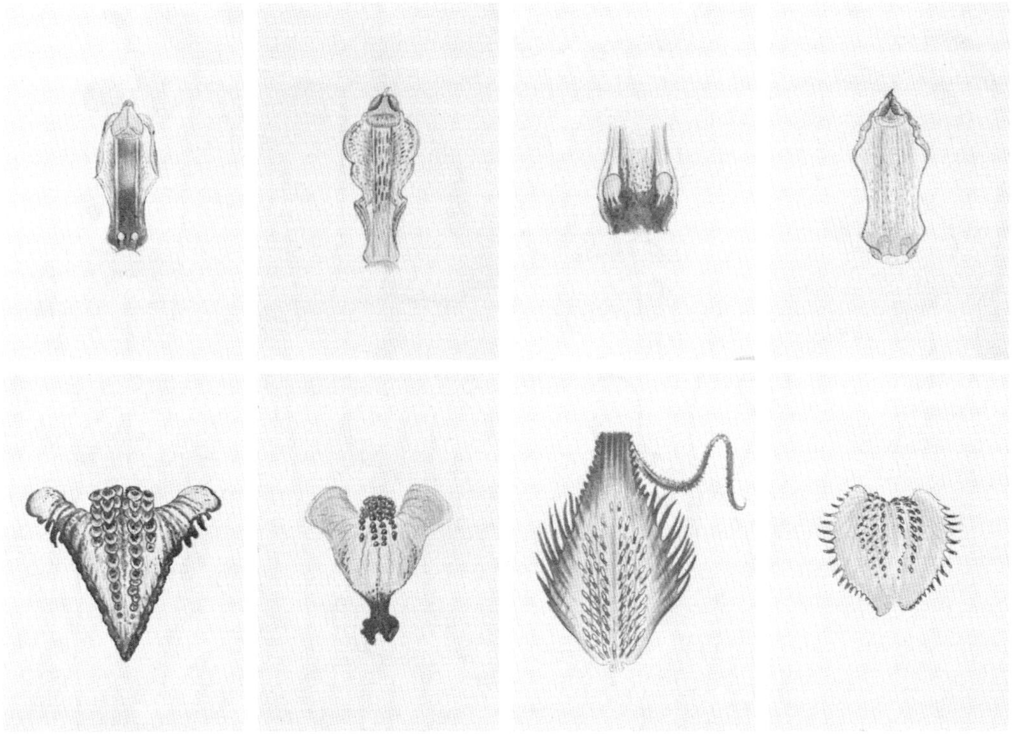

Variation in the size of false wasp 'eyes' at the base of the column in four spider caladenias but *Caladenia integra* has no eyes at all. Now compare the dummy wasp bodies on the lip petals and the variation in lip length, shape, margins, calli size/shape and calli numbers. From left to right: *C. tesselata*, *C. integra*, *C. echidnachila* and *C. pallida* (taxonomy follows Nicholls 1969). These images are modified from W.H. Nicholls' *Orchids of Australia* 1969.

in contrast to their lip petal. You can't miss them. Under a little more magnification, you will notice that some species display separate, slender calli with hooked tips. In other flowers they may be packed together, with flattened, overlapping heads, blackish but lustrous like a cluster of currants drying on the vine. The margins of the lip petals of some spider caladenias also display gracefully, curved teeth like a Spanish comb.

Most flowers of spider caladenias have three, elongated sepals and two shorter, narrowing lateral petals. If you compare how different species display these floral parts, you would conclude that these orchids have a distinct sense of posture. Sepals may spread out stiffly like arms of a sea star, dangle limply underneath the lip or curve upward making the flower look like it's lying on its back like a dead horse with rigor mortis. On breezy days the positions of the long, stiff and leg-like sepals of some flowers make them look like they are pirouetting or galloping over the heath. Stripes on the sepals and splotches on the lip boldly announce their presence in reds, yellows, greens and purples. Now check out the tips of those sepals. Note that one to all three of them are swollen and club-like. They are scent glands (osmophores), as identified in Chapter 1, and experiments show these knobs definitely secrete something that attracts male wasps.

Does this extraordinary diversity of floral presentation mean that every spider caladenia in Australia is pollinated by only one species of wasp? It's unlikely, as it's estimated there are at least 1,000 species of thynnids in Australia. For one thing, these wasps often make mistakes but this has attractive consequences. Recently, 18 'rare' *Caladenia* species had to be reclassified as recurrent hybrids especially in Western Australia where dozens of different species of spider orchids coexist in mixed colonies and have overlapping flowering periods. The best books on Australian orchids warn about spider orchid promiscuity. For example, the drooping spider orchid (*Caladenia radialis*) is recorded as forming hybrids with at least half a dozen other species including the bluebeard (*Caladenia deformis*).

Obviously, male thynnids are either more fallible or less faithful than we thought they were, but there could be important consequences to these hybrid swarms. The sheer frequency of reported hybrids suggests that evolution is ongoing and rapid in this orchid lineage. We are looking at incomplete segregation between species because isolation barriers between diverging populations have not had time to evolve. Plant hybridisation is not always a waste of offspring. While first generation hybrids do not commonly evolve into new species *de novo*, they represent living gene banks on which one or both of their parent species may draw as wasps continue to make mistakes each generation changing frequencies of genetic variation in one or both orchid populations. Over many years, some beneficial genes from one species of spider orchid may 'filter' into another bettering its chances to change and expand its distribution. We call this process introgression.

The more we study the spider orchids the more we expand our ideas on how they manipulate male wasps. The evidence suggests there are at least three 'strategies'. First, species with black, clustered and often hairy calli exploit lovesick males. Some of these flowers take the deception one step further adding pairs of fake eyes under the column as we saw in the bearded orchids.

Second, a strong sexual deceit doesn't apply to all those pretty white or purple females dancing in spring bushlands that I've admired for decades. Some like the scented spider orchid (*C. fragrantissima*) offer perfumes a human can detect and enjoy. These flowers are not luring wasps with the promise of sex. They are employing much the same modes of advertisement as do the hundreds of unrelated wildflowers, shrubs and trees around them that actually offer nectar and pollen rewards. The difference is that some of the prettiest spider caladenias are again making false promises. There's no nectar between the base of the lip petal and the column. The wasp who switches his allegiance from the nectar-filled cups of a Geraldton wax flower (*Chamelaucium*) or tea-tree (*Leptospermum*) to one of those larger, more extravagant looking spider orchids is in for a disappointment. If the same wasp is foolish enough to visit more than one orchid in good faith, then cross-pollination can occur.

Didn't I mention three strategies? Back in 2008, Dr Kingsley Dixon invited me to view his potted collection of Australian orchids while he was the Director of Research at the Kings Park and Botanic Garden in Perth, Western Australia. His collection of spider orchids was blooming and he pointed out something I never expected to see. The lip petals of some species bore sticky beads of nectar. I tasted them and they were sweet. In fact, some droplets appeared to be produced at the tips of those elegant comb teeth on the margins of the lip petals. The sight of those beautiful petals oozing nectar brought back memories of the flowers of those homely Brown's peonies gushing with drinks for hairy flies and social wasps in Chapter 6. The swamp spider (*Caladenia paludosa*) and stiff spider (*C. rigida*) orchids are reported to secrete nectar, and male wasps are not their only visitors. Small bees and hover flies are known to transfer their pollinaria. Both of these spider caladenias are known to hybridise with other orchids so they could be passing on genes that control sites of liquid secretion to future generations of new species.

The multiple ways in which Chinese and Australian orchids manipulate unlikely pollinators, like wasps, must tell us something very important about the success of this plant family. New orchid species are still described every month in botanical journals as expeditions forge trails through under explored mountains and forests. Orchids are usually small but durable plants colonising tree branches and rocky outcrops and they will continue to cling to microhabitats as humans degrade their environments. Within the past few years, botanists at the Missouri Botanical Garden have even added new orchid species to the flora of Madagascar, despite the fact that so much of the island's native vegetation has been destroyed over centuries by deforestation and soil erosion.

When we play the numbers game, this means that in some parts of the world (including Australia) one out of every 10 species of plants is an orchid. The basic architecture of their flowers, combined with their ability to produce a range of scents, shapes, colours and nectars, indicates that different orchid species survive to reproduce because they are exploiting quite a wide range of pollinators, including wasps and flies. They will continue to do so given sufficient time and access to protected and stable landscapes.

We must remind ourselves, again, that orchids aren't the only flowers exploiting unlikely pollinators. The female thynnid orchids in this chapter exploit the larvae of beetles. Do flowers exploit adult beetles as pollinators? To answer the question, we need to visit South Africa.

# 8

# Beauty through a beetle's eyes

*And when the Beetle woke up he crept out and looked around him. What splendour there was in the greenhouse! ... what a riot of green, and blooming flowers, red as fire, yellow as amber, or white as freshly fallen snow! 'What magnificent plants! How delicious they'll taste when they're nice and decayed!' said the Beetle.*

– The Beetle, *Hans Christian Andersen. Translated by Hean Jersholt.*

DURING 15 YEARS OF LABORATORY WORK IN SAINT LOUIS MY RESEARCH partner was Dr Peter Goldblatt, the Curator of African Botany at the Missouri Botanical Garden. Peter brought me such interesting specimens from his flowers blooming in the Republic of South Africa. They were never gifts and, once I finished analysing them, they were donated to appropriate museums. There were tangle vein flies (Nemestrinidae) with tongues twice the lengths of their bodies and even a few oil-collecting bees (*Rediviva*) whose squeegee-like forelegs resembled the scythes of a praying mantis.

It was my job to check each insect for the pollen of the flower on which it was caught. This must always be done gently as pinned specimens grow increasingly brittle the longer they are exposed to the air. Each impaled corpse was placed on a glass slide and 'bathed' in a few drops of solute to flush off any particles they carried. The body was removed, air-dried, then placed back in its collection box. Once the liquid on the slide evaporated, any grains left on the glass were stained with a basic fuchsin dye before topping them off with a protective coverslip. The slides rested overnight and then I could view them under a microscope. By then, the walls of the pollen grains were stained pink making it much easier to find the characteristic wrinkles, spines, pores, pore caps and/or slits that allowed me to identify members of the iris family (Irdiaceae) versus those from the daisy (Asteraceae) or protea (Proteaceae) families.

Most people do not realise that the outer wall of most pollen grains are colourless. The yellow, orange or rusty-red blobs or smears we see are droplets of a natural vegetable oil coating the grains. The oil is made inside each anther and it coats the grains before the anther opens exposing them to pollinators. This greasy coat helps grains stick to each other and adhere to the bodies of visiting animals. As a rule, the anthers of wind-pollinated plants make almost no oily coat and explains why the 'powder' released by grasses or male pine cones appears white to our eyes. Eventually, the greasy and pigmented droplets associated with animal-pollinated flowers recede into the caves and crevices on the pollen grain wall or vanish altogether the longer they cling to surfaces outside the flower that made them. You won't see many oil droplets under a microscope if the pollen was taken from an insect pinned weeks or months ago, or from a dried, pressed flower.

## WHAT MONKEY BEETLES DO AND DON'T DO

I distinctly remember the day Dr Goldblatt handed me my first box of beetles from South Africa. There were rows of pinned, chunky and bristly bodies. Some had a metallic sheen to their shells and resembled tanks badly camouflaged with armpit hair. I learned that South African naturalists called them monkey beetles and entomologists placed them in the tribe Hoplini in the family Scarabaeidae. At least 65% of all the world's monkey beetles (over 1,000 species) live in South Africa. These insects are further segregated into more than 50 genera, such as *Anisonyx*, *Clania*, *Lepithrix*, *Lepisia*, *Pachycnema* and *Peritrichia.* I washed hundreds of monkey beetle corpses over the next decade and shook hands with visitors with my dye-stained fingers.

You might ask, aren't scarabs the dung beetles of ancient Egypt? Why would any of them visit flowers or wear pollen? The sacred scarab (*Scarabaeus sacer*) of the Pharaohs does belong to this family but so do more than 30,000 species and most don't lay their eggs in rolled balls of dung. These beetles are more likely to lay their eggs in soil or rotting wood and then their hatching larvae dig down and eat plant roots. We often call them curl grubs at that stage. As winged adults, many continue to live on strictly vegan diets. June beetles (*Cotinis*), for example, prefer to eat ripe fruit while other species are observed as they chew the edges off fresh leaves and flower petals. We often call them chafers.

Some plant scarabs are regarded as noxious pests, especially when they are introduced to other parts of the world by accident. Japanese beetles (*Popillia japonica*) arrived in America shortly after the first decade of the 20th century. Now they are found in all but a few states in the American West. They are handsome as adults in their green and bronze armour, but a gardener

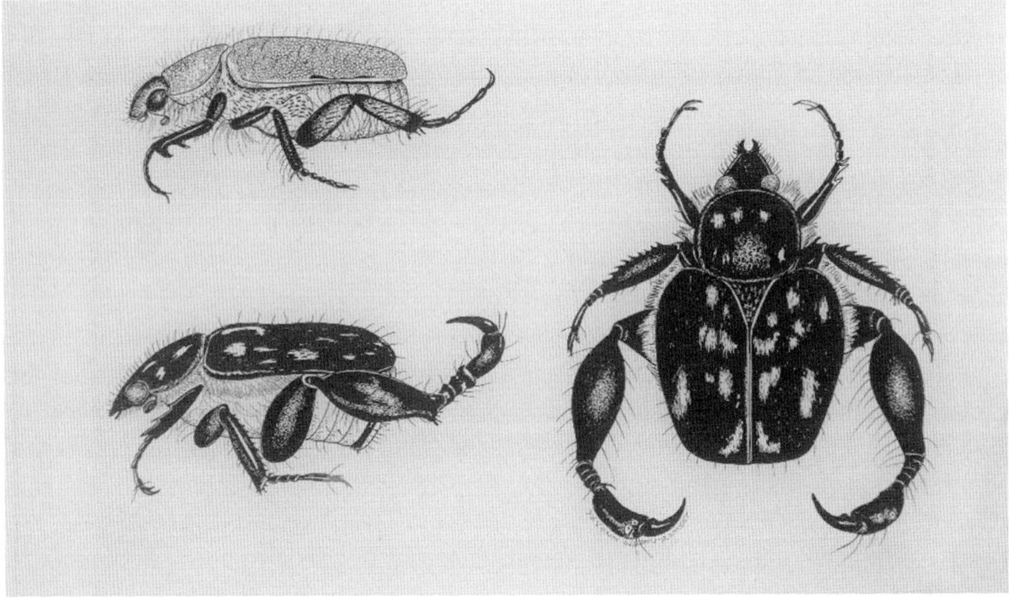

*Lepisia rupicola* (top) and *Pachynema crassipes* (bottom left and right) are two of many species of South African flower scarabs. Their swollen limb joints remind me of a hairier and fatter version of Popeye the Sailor. Illustration by Y Wilson-Ramsey.

will reach for an insecticide instead of a camera when they chafe their prize roses. Agricultural and horticultural agencies keep lists of the leaves of hundreds of wild and domesticated plants skeletonised each year by Japanese beetles. They even punch holes in thistles (*Cirsium*) and poison ivy (*Toxicodendron*) with impunity. Their curl grubs live primarily on a diet of grass roots turning stretches of lawns yellow or bare.

Monkey beetles, however, represent a smaller, evolutionary step from destroying a plant's blossoms and fruits to a less destructive diet of pollen and nectar. Working with Dr Goldblatt, I learned about the long-term and profound effect these insects had on the diversity of many unrelated families of plants in southern Africa. In our largest and longest study back in the 1990s, we found that at least 42 beetle species either shared their flowers peaceably with flies and bees, or they proved to be the sole pollinators of dozens of wildflowers representing six or seven different plant families. Many of these beetle-dependent plants grew as perennial bulbs relegated to regions with wet, mild winters and hot, dry summers. This included members of the asparagus (Asparagaceae) and goldstar (Hypoxidaceae) families, but certain members of the iris family (Iridaceae) were especially dependent on beetle pollinators. Other botanists, often inspired by our work, found monkey beetles pollinating flowers of members of the native daisies (Asteraceae), ice plants (Aizoaceae) and even the carnivorous sundews (Droseraceae). The trapping tentacles of South African sundews were not strong enough to catch and eat their pollinators.

As monkey beetles lack the elongated tongues of bees and tangle vein flies, the flowers of South Africa dependent on these insects, regardless of family, had similar shapes. Most flowers pollinated primarily by monkey beetles offer edible pollen in shallow cups or bowls. The rims of these blossoms are wide enough to accommodate more than one visitor at the same time. This turns some flowers or flower clusters into very busy places where food is not the only issue. The flowers preferred by monkey beetles are a place to eat, meet and mate. Like so many other insects, a pair of beetles may remain *en copula* for hours. The female grazes on the pollen in the anthers while the mounting male remains in place blocking further access to the female's genital opening. The longer he remains in this position the more likely his sperm and his alone will sire the next generation of grubs. It's advantageous, then, for some male beetles to, at least temporarily, turn fresh blossoms into personal territories. They won't bother bees or butterflies, but male monkey beetles can become aggressive. Some species have a 'push-me-pull-you' behaviour that includes turning a rival over on his back.

I watched this combination of jousting and sumo wrestling in October 2024, as a guest of the famous Kirstenbosch National Botanical Garden in Cape Town, South Africa. The broad heads of native members of the daisy family were designated sports arenas. The females fed or froze in place while the males bumped and shoved each other.

In contrast to so many of the flowers I've introduced in the previous chapters, Drs Goldblatt and John Manning found that most of the ones visited exclusively by monkey beetles lacked discernible scents and were not dripping with lots of nectar. Pollen was often the only item on the menu, but pollen is never easy to digest. Once again, the outer wall of each grain is based on a chain of hard to break polymers making it tougher than some man-made plastics. One of the simplest ways to access the nutrients inside the grain, as practiced by bees, hover flies and some bats, is to drink sugary nectar first before swallowing pollen. Once the ingested grains absorb the fluid through their pores or slits they swell until they rupture releasing food directly into the insect's gut.

Flower beetles in other parts of the world have additional ways of extracting the nourishment guarded within the pollen wall. Some have a pair of pollen-cracking 'molars' on their mouthparts. One tropical species living on the pollen of some palms is reputed to first swallow the hard, protective 'stone cells' on the skin of its favourite palm flowers. This grit allows it to wear down the pollen walls like a chicken's gizzard grinds down seeds.

Monkey beetles appear to lack these adaptations. Shelley Johnson and Susan Nicolson at the University of Cape Town examined pollen digestion in the monkey beetles *Peritrichia cinerea* and *Pachynema flavolineata* using several techniques to see what happened to grains in the mouthparts and through the guts of these insects. They even used a series of different stains to predict where the digestive systems of the beetles picked up morsels of starch, protein or fat. Monkey beetles are not biters. Each pair of lower jaws (maxillae) on opposite sides of each beetle's face are modified to form hairy brushes sweeping whole grains down their throats without smashing the pollen walls. In the absence of drinks of flower nectar or the consumption of grit, their digestive systems still performed without complaint. Their dung consisted solely of empty and deflated pollen shells. The authors suggested that these insects have one or more digestive enzymes that enter the indigestible, outer pollen wall through each grain's pores or slits.

## SIGHT NOT SMELL

What about the relative absence of scent in the monkey beetle-pollinated flowers described in the original field notes of Goldblatt and Manning? Beetle-pollinated flowers made by members of the iris family in South Africa are odourless compared to their closest, strongly scented relatives pollinated by bees, hawkmoths or dung flies. For example, freesias, members of the iris family (Iridaceae) which are native to South Africa, are prized garden plants in temperate countries. In addition, horticulturists have hybridised freesias for garden beds and potted displays because they smell so good. Following a visit to South Africa, biochemist Roman Kaiser never had any problems coaxing hundreds of compounds out of wild species of *Freesia* (Chapter 1) and/or from other members of the iris family including species of *Gladiolus*, *Hesperantha* or *Ferraria.*

In fact, scent poor beetle-pollinated flowers represent a contradiction in natural history because flowers pollinated primarily by beetles are expected to smell strongly according to comparative studies beginning in the first decades of the 20th century. Some people described the odours of beetle-pollinated flowers as reminiscent of fermentation or luscious ripe fruits. In particular, they remarked on the heavy scents of the blossoms of beetle-pollinated species of magnolias, tropical water lilies and many members of the philodendron (Araceae) and soursop (Annonaceae) families. The same botanists had little to say about the flower colours of these wild plants. With important exceptions, the smellier beetle flowers tended to be a pale and a monochromatic white, pink or creamy-yellow. A few are decorated with irregularly spaced purple or brownish blotches.

Compare these descriptions to the riot of colours displayed by flowers that attract monkey beetles in South Africa. Unlike all the plant species I described in the previous fly and wasp chapters, and will again in Chapter 10, flowers preferred by monkey beetles are far more vivid. Observations and field experiments indicate that monkey beetles are attracted most often to reds, oranges and yellows. Kim Steiner of the Compton Herbarium, Claremont, South Africa, put out traps of artificial flowers of three different colours for the beetle *Lepithrix hilaris,* which she observed foraging on eight local wildflowers. Sixty beetles ventured into the orange trap, more than 35 were found in the yellow trap, but less than 10 were found in the white trap. In South Africa, different species of monkey beetles fancy different colour patterns. Unlike flower beetles in other parts of the world, they depend on what they see rather than what they smell.

## BEETLE MARKS, A SHARED ADVERTISEMENT

There are some white beetle-pollinated flowers in southern Africa as well as blue and purple ones, but most share an important feature with those that have red, orange and yellow petals. They are marked with a contrasting sooty or shiny pattern that looks black to human eyes. Sometimes the black mark is speckled or feathery, and is edged or splashed with white and other colours. Botanists have come to call these patterns beetle marks. At one time, entomologists thought that beetle eyes lacked depth of focus and the sociable but aggressive insects mistook the marks for more beetles. A flower bearing beetle marks was compared to a duck hunter setting out decoys on a lake. Botanists presumed that the more beetles bumped around together on a flower, the more likely they'd contact anthers then transfer grains to receptive pistil tips on the next flower they visited. The marks then were thought to give each plant an adaptive advantage as a rendezvous flower stimulating the natural behaviour of male and female beetles to look for the best places to eat and mate. It's not that simple. Field tests showed that certain beetles were just as likely to visit coloured models of flowers without beetle marks as they did with beetle marks. Depending on the species of African wildflower, beetle marks can be unusually elaborate and intricate, but their full significance is still not understood.

What beetle marks definitely do is encourage people to admire, grow, breed and buy the plants that have them. Go to almost any garden centre or nursery in good weather and find the hybrids of the popular South African Barberton or Transvaal daisies (*Gerbera*) with their centres of blackish disc flowers surrounded by ray flowers of many colours. Cultivated forms of gazania or treasure flowers (*Gazania*) are even more popular as some of the most ornate and complicated beetle marks decorate the base of each of the long ray flowers that make up each composite flowering head. If you must buy a daisy in a nursery wouldn't you rather take one home in which the flowers are 'freaked by jet', to borrow a phrase from Milton?

And yet, these daisies are not the most flamboyant exploiters of monkey beetles to come from South Africa. That award belongs to a few members of the iris family, specifically members of the genus *Moraea.* Yes, most *Moraea* species continue to offer some scents and a nectar reward and are dependent for pollination on a variety of native bees including a local race of the African honeybee. However, there has also been a dramatic shift in the floral evolution of at least eight *Moraea* species, and all eight are preferred by monkey beetles. Online seed and bulb catalogues offering these plants, or their hybrids, are likely to entice their customers with words like 'plant porn', 'must haves' and 'spectacular'. Presentation of their beetle marks vary as black dots or delicate brush strokes become engulfed in layers of yellow, or orange. Two species, the peacock moraea (*M. villosa*) and the rooiflappie (*M. tulbaghensis*), out do all the rest. Their marks glitter as they incorporate truly iridescent blue patches.

Drs Golblatt and Manning found that at least eight beetle species visited wild populations of rooiflappie and peacock moraeas pollinating their flowers. My analyses of pollen carried by these insects expanded the story. While these beetles obviously preferred cup- or bowl-shaped flowers with contrasting colours, their loyalty to any wildflower was limited. It didn't matter whether the beetle was collected on a peacock moraea, sundew, goldstar or gazania. Most of the pinned specimens I washed carried the grains of more than one wildflower species.

## THE INTERDEPENDENCE OF WILDFLOWERS AND MONKEY BEETLES

If most monkey beetles have their own shopping preferences for flowers with particular colour combinations, but only a limited sense of customer loyalty, that must impact different species of wildflowers blooming only a few footsteps from each other. Ecologists would say that beetle-dependent plants in the same South African community are members of the same guild because they are all exploiting the same environmental resource (monkey beetles) in much the same way. Think about it. We've already considered two pollination guilds in Australia. There is the guild of hundreds of winter orchid species all dependent on fungus gnats and the guild of hundreds of species of spring-flowering, bushland orchids visited only by male wasps. The difference in South Africa is that the different plants blooming together in the same habitat may not be closely related to each other, but they may still be forced to share the same, closely related beetle species as their primary pollinators.

While many of us think that Nature is far too generous with the insects she produces each growing season we must now agree that pollinators can be as limited a resource in a plant's lifecycle as the nutrient levels in its soil or annual rainfall. As plant species in the monkey beetle-pollinated guilds of South Africa must share some of the same beetle species, the flower colours displayed by completely unrelated wildflowers tend to converge. As a limited number of grubs survive each year to emerge as winged adults, different plants must also compete for their limited visits. Natural selection should favour flowers that attract the most eager pollinators. In South Africa, it pushes the evolution of visual attractants to gaudy extremes.

Now let's look at it from the beetle's point of view. The amount of pollen produced by any flower is also a limited resource. If all beetles go to the exact same flowers won't some of them starve? What prevents unfair and disastrous competition between beetles of different sizes belonging to so many different genera?

Two entomologists, MD Picker and JJ Midgley, set up a series of colour traps at three sites in the Biedouw Valley and at Darling, South Africa. The traps were large, transparent bowls each painted a different colour and each one was filled with 2 cm of water. Insects tempted into visiting these big decoys drowned and could be easily collected and identified. The two entomologists also recorded which insects visited the most easily identified target flowers in bloom at the same time. They found that no beetles liked the colour green and all avoided the transparent bowls used as experimental controls. Dishes daubed red, yellow or orange were preferred by the fuzzier species of *Lepithrix* and these beetles were observed to prefer wildflowers with the same range of colour patterns. One species, *Gymnoloma tibialis*, proved to be an extreme specialist visiting only yellow bowls and yellow flowers. It became clear that different beetle species appreciated different pigmentation patterns and different arrangements of wildflowers on their stalks. Blue and white dishes were most likely to attract *Anisonyx* and *Peritrichia* species and these insects were found, in turn on the blue and white flowers of members of the mint family (Lamiaceae). The ice plants and daisies were far more likely to be visited by smaller and comparatively hairless beetles belonging to species of *Pachynema* and *Heterochelus*.

If different beetle species have different colour preferences natural selection may favour a plant species where colour variation is written into its genome. The more colour forms offered by the same population, the greater the diversity of beetle pollinators attracted during the same flowering season. With each generation, the frequency and intensity of each flower colour must hinge, at least in part, on how attractive their colourful parents were to previous generations of beetles. You can see the man-made impact in a florist's shop or nursery centre. One of the reasons we have so many garden varieties based on the Barberton daisy (*Gerbera jamesonii*) of South Africa is due to the distribution of colour forms in its wild populations. While orange-red flowering heads dominate most sites, some isolated populations produce and add less common yellow, white and pink forms.

## WHY FLOWER COLOURS VARY AMONG POPULATIONS OF POPPY SUNDEWS

Consequently, the same species of beetle-pollinated wildflower in South Africa can break up into populations where different colours occur at different frequencies across its natural range. So called, colour shifts may be due, in part, to the density and diversity of resident beetles versus the diversity and density of other kinds of wildflowers dependent on the same beetles. The best-studied case is the poppy-flowered sundew (*Drosera cistiflora*) of the Greater Cape. It shows at least five possible colour forms as recorded by Steve Johnson, Caitlin G. von Witt and Bruce Anderson. All of the colour forms lack nectar and scent but they all have a smudged or blackened centre, as usual. Pink and white forms are most common throughout its range but the South African scientists found six populations dominated by yellow flowers, nine by red, and four by purple.

The scientists observed 16 sundew populations across 680 km and wondered whether petal pigments were really influenced by soil and rock chemistry, not genetics. In some parts of the world, rainwater percolating through different soils becomes acidic or alkaline depending on the underlying geology. Roots take up the solution altering petal colour. This was not the case for the sundews. Their populations grew on 21 different types of substrates and no colour form could be matched to a specific kind of soil or bedrock. For example, pink, white and purple forms all grew together regardless of whether they were rooted in loam or sand. When plants of known colour forms, from known sites, were collected and grown on in pots of different soil types their petal pigmentations didn't change.

What did change from site to site was the diversity of 12 species of monkey beetles and other insects, and so did their pursuit of differently coloured flowers. Almost 90% of the visitors to white or red flowers were all identified as the same monkey beetle, *Lepisia rupicola*. These insects also showed a minor partiality to pink blooms. In comparison, almost half of the pollen eaters on yellow flowers were identified as the same species of *Lepithrix* beetle. *Lepisia rupicola*, in comparison, never perched on yellow petals. More than 60% of the visitors to purple flowers were beetles in the same species of *Omocrates*, while *Lepithrix* species and *Lepisia rupicola* wouldn't be caught dead on purple. As each species of monkey beetle doesn't show the same broad distribution as the poppy-flowered sundew, regional variation in the plant's petal colours must be of some benefit. It suggests that different beetle species are more likely to visit and pollinate different colour forms when these insects are restricted locally to isolated populations of sundews.

The petals of all the colour forms of the poppy sundew shared one more thing in common. None of them reflected ultraviolet light. It's sobering to think that monkey beetles may see much the same range of colours as we do. Compare that to most studies on social bees. They tend to disdain reds and oranges yet their compound eyes are sensitive to UV patterns etched genetically into the skin cells on flowers which remain invisible to our eyes.

## WHEN ORCHIDS CHEAT ON BEETLES

With such a broad and diverse guild of beetle-pollinated plants in South Africa we should ask: Do all the plant species offer edible pollen or is there room for a cheater or two? The immediate answer is yes, and as usual an orchid is exploiting the system. Kim Steiner looked at *Ceratandra grandiflora*. There are five more orchid species in the genus *Ceratandra* but their flowers all smell of aliphatic and ketone compounds. Each has a lip petal bearing an oil-secreting gland. These fatty secretions are collected exclusively by oil-collecting *Rediviva* bees, as mentioned in the opening paragraph. These bees feed the orchid's grease to their larvae. In exchange, the bees transport the pollinaria of the orchids, cross-pollinating flowers as they forage from plant to plant.

A flower of *Ceratandra grandiflora*, though, has no scent and the oil gland on its lip petal is usually reduced to a nub, so it is usually unable to secrete any oil at all. Bees avoid this orchid although each plant presents its flowers in tight, cheerful, yellow clusters. Despite the bee boycott, a quarter of the long-lived flowers on a single stalk are pollinated within the same day. The only visitors are two species of monkey beetles and both show their greatest periods of activity around mid-morning. Both insects prefer older clusters of flowers and neither species of monkey beetle attempts to sample the reduced glands on the lip. As they inspect and crawl about each yellow cluster, they stumble across the sexual columns in each flower triggering the release of pollinia onto their leg joints.

Male beetles appear more interested in the orchids than females and patrol the flowering heads looking for potential mates. When they can't find a suitable mate after a minute they fly to a second plant in bloom unconsciously spreading the pollinaria already attached to their legs. The beetles do not receive much in return. Dr Steiner observed only one female of *Heterochelus podagricus* on one orchid cluster and she mated with more than one male present. In the absence of edible pollen, no beetles starve to death as there are plenty of other beetle-pollinated species in bloom at the same time. Duped by orchids for an hour or two at most, these beetles quit and move on to a brunch served by yellow goldstars, gerbera daisies, treasure flowers and other species.

The interdependence of native plants and monkey beetles in South Africa seems almost paradoxical. It appears to be based on genes that retain the simplest flower shapes and invest heavily in colour variation, but then they turn off glands that would manufacture scent, nectar or oil. This system is so cheap and effective that lineages of entirely unrelated plants invest in it. It's a dynamic and ongoing system undergoing relatively recent changes when we consider that the yellow *Ceratandra* orchid is still carrying around the nub of a gland its ancestors once used to reward bees.

Some would argue that the wildflowers of South Africa have domesticated monkey beetles. As winged adults they no longer represent a threat to flowers. What determines whether an animal aids a plant's reproduction, thereby increasing the plant's fitness, or contributes to the destruction of its seeds, thereby reducing the same plant's fitness? What can happen to plant populations over several generations beset by visitors capable of varying degrees of destruction? Specifically, can the relationship between a plant and its pests evolve into a mutualistic alliance? To look at one emerging example we need to return to some of the most beloved flowers in Australia with a side trip to some of the drier areas of North America. Warning, I'm about to become a little sentimental.

# 9

# Native rose and its invisible worm

*O Rose thou art sick.*
*The invisible worm,*
*That flies in the night*
*In the howling storm:*

*Has found out thy bed*
*Of crimson joy:*
*And his dark secret love*
*Does thy life destroy.*

– The Sick Rose, *William Blake*

MOST OF US DO NOT WANT TO BE REMINDED THAT WE LIVE IN A WORLD FILLED with cocoons and caterpillars. Entomologists, farmers and those employed in the food and fibre industries do not need reminding either and they take a pragmatic approach. Zoologists note there are close to 200,000 species of Lepidoptera (butterflies and moths) and they regard it as the second largest order of animals on the planet. It is second in number of species only to beetles in the order Coleoptera, as in Chapter 8. As adults most lepidopterans bear hairs flattened into removable, overlapping scales on their bodies and wings. Most have mouthparts that interconnect to form a feeding tube to take in fluids. Most are classified as moths, while our beloved butterflies make up only 5–7% of described species in the order. It's fair to say we live in a moth-eaten world.

Farmers are more likely to suffer than profit from moths, with the exception of a single domesticated species, *Bombyx mori*, still under domestication after more than 5,000 years. Called a silkworm in English the animal is not a source of expensive filaments until after it has completed its cocoon and has abandoned its caterpillar form in favour of an immobile pupal stage. In contrast, there are hundreds of species that devour and ruin crops throughout their larval stages, and they are often very specific. The cutworms and armyworms that invade vegetable gardens and lawns gnawing through the seedling stems belong to the huge family of owlet moths (Noctuidae). Some owlets destroy a wide variety of useful plants, while others have specialised tastes eating only grasses, turnips or lettuce, and so on.

The caterpillars of different moth families share one thing in common, they tend to devour what we like best. Many attack specific organs on a plant as it matures, such as fruits or seeds. The family of codling moths (Tortricidae), resident on all continents except Antarctica, are literally the worms in our apples, plums, peas, peaches, grapes, pomegranates, kiwis and more.

I belonged to that generation of American children taught never to purchase an ear of sweet corn without first pulling back the leafy shucks to check the cob for kernels ruined by corn borers (*Ostrinia nubilalis*) in the family Crambidae. I still do so at farmer's markets and roadside stands, and suffer the angry glares of their proprietors.

Cereal crops are not safe from moth attack after grains are harvested and milled into flour either. The same goes for our dried fruits and shelled nuts. There are plenty of pantry pests in the genera *Plodi*a, *Ephesti*a and *Cadra* belonging to the family Pyralidae. When something small and drab flutters out of your kitchen cabinets it's a sign you have not sealed your jars well and you are nourishing moth offspring.

Finally, there are the remarkable caterpillars maturing in animal-based fibres and leather. Holes in expensive textiles including heirloom carpets and tapestries indicate that those cedar balls and naphthalene crystals were applied incorrectly, or didn't do their jobs during summer storage. These insects have moved in on us over thousands of years. Civilisation offers them larger and more dependable caterpillar creches compared to feathered or fur-lined nests or the rotting skins of animal corpses. You must admit that completing a lifecycle in a wardrobe, in the absence of free water or a well-hydrated food source, is an extreme way to provide for your offspring. Most clothes moths belong to the family Tineidae and are entirely dependent on protein-based fibres. Their larvae are no threat to plant-based threads based on cellulose, such as cotton or linen.

What do all these insects have in common aside from depriving us of clean food and ruining our possessions? Most go unnoticed until it's too late because the winged parent is so small and dingy. No one ever says, 'Look at those beautiful moths laying eggs on my angora sweater'. What should also be obvious is that, while these insects have an expensive impact on the human economy, we must also expect they play pivotal and complex roles in food webs on a worldwide scale. While some destroy a plant's ability to grow and reproduce, many plants would be unable to reproduce without moths.

## MOTHS AS POLLINATORS AND HOW ANCIENT?

To recap what was described in Chapter 1, Knut Faegri and Leendert van der Pijl reviewed and itemised the range of floral shapes, attractants and edible rewards offered by hundreds of different species of flowering plants. They realised that a plant's genetic base determined how and when it presented itself to animals sharing the same environment. Colour, scent, and the nutritional value and volume of awards paralleled the season and time of day when flower buds opened. These two botanists became curious about flowers pollinated almost entirely by moths and this drew them to plants that waited to open at sundown and/or into the night.

Of course, not all moths, including the silkworm and its relatives, feed as winged adults. Most that do retain a coiled proboscis to probe for nectar. Faegri and van der Pijl compared the flowers of different species and noticed a convergence based on whether the moths settled on the flowers or hovered in mid-air while they drank. They found that moth-pollinated flowers varied in size but usually took the form of tubes and funnels with nectar accumulating at the base. With such narrow openings a hovering moth couldn't push its broad, beating wings very far into a flower and pollen was usually dotted onto its head or at the base of its proboscis. In comparison, weak-flying or settling moths were more likely to land on flowers. Once perched on the rims of lots of tiny, congested floral tubes, pollen was deposited on their legs and undersides. In either case, flowers pollinated by moths lacked vivid colour patterns and looked white, pale yellow or greenish to human eyes. While some white flowers may reflect the light of the moon, visibility is obviously a problem in the dark. Flowers for nocturnal moths signalled their presence with strong scents some gardeners and perfumers found pungent but pleasant. That explains the popular culture of tuberoses, jasmines and ornamental tobaccos, as well as the large but short-lived flowers of some night-blooming cactuses grown by fanatical hobbyists.

Night-blooming flowers for moths sing the usual song to prospective pollinators. They offer nourishment in exchange for the successful transfer of sperm, but the system of attractants and rewards is played as a nocturne. Scent may be just a little more important than sight in this syndrome, but is that the whole story?

Wing scales are a signature feature of the Lepidoptera. They fossilise and evidence places their origin back into the Jurassic Period from 260 to 190 million years ago depending on which scientists are willing to confirm that the specimens they've examined belongs to an ancestral moth or a caddis fly in the related order Trichoptera. It means, though, that moths are probably older than the earliest true flowers indicating that the first proboscis had other functions, such as sucking up water droplets in dry zones. It also suggests that moths may have owned the night sky for millions of years until the evolution of a hot-blooded and ravenous group of mammals we call bats (Chapter 10). The earliest bat fossils are only 51–55 million years old placing them within the Eocene. No one is sure where they first evolved as their bony wings encouraged their migration through the continents of the Northern Hemisphere and south into Australia and Africa within a surprisingly short period of time.

Besieged by sharp-toothed, echolocating bats, natural selection favoured those moths that survived nightly attacks with new adaptations and some may have done so by vacating their nocturnal niches. Many moths from different families, as we will see, have come out of the night closet. Sometimes members of the same family share diurnal, crepuscular, and nocturnal cousins. While working in the Heng Duan Mountains of Yunnan between 3,000 and 3,300 m a.s.l., I was impressed by the stratified, 24-h cycle of sphingid moths. As the sun set, the greyish and brown, night-flying sphinxes visited the strongly scented flowers of the rein orchids (*Habenaria*) related to the blunt-leaved orchid (*Platanthera obtusata*; of Chapter 3). By day, a second set of smaller, faster sphinxes with clear wings found nectar in the wet meadows foraging in dense populations of pink and purple primroses.

We know that certain lineages of moths evolved over and over to lay eggs in fruits and seeds. We also know that different ancestors of modern moths exiled themselves to the daytime. Do these two life choices ever tie together in Australia and in other countries? I think they do but it has taken me over 45 years to understand some of the evolutionary implications.

## MY BORONIAS, A SENTIMENTAL JOURNEY

I arrived in Melbourne at the end of August 1977 in time for my first austral spring. Urban Australia, as I soon learned, smelled good. It was a great time for horticultural introductions especially at the University of Melbourne and the Royal Botanic Gardens Victoria. The beds were filled with preferred heirloom cultivars imported from Eurasia, but then you'd cross a lawn and look at attempts to tame and grow native plants. Most of the native species were small trees and shrubs. The scents in my nose shifted from old rose breeds and clumps of sweet violets (*Viola odorata*) to the musty, almost meat-like smells of banksia cobs.

Drawing of brown boronia (*Boronia megastigma*). Adapted from Curtis's Botanical magazine, 1801.

Contrasts seemed greatest around the School of Botany at the University of Melbourne. Professor TC Chambers (1930–2023) was involved directly with this marriage of native and British favourites. In the System Garden at the rear of the Botany Building, goldfish in shallow lily pools were caught and eaten by resident kookaburras. The lawn was seeded with tiny English daisies (*Bellis perennis*) and fruit trees were espaliered to grow flat against the wall of the seminar room. In front of the building, by contrast, there was the vegetation you'd see in the gullies of the Dandenongs. The building's shade benefitted the growth of native fern trees, a speciality of Professor Chambers.

By mid-September, I found myself drawn to a bed reserved for plants preferring more sunlight. There was a spindly little shrub festooned with curious, bell-like flowers. Each flower bore four curved petals that were a deep reddish-brown on the outside and the most cheerful canary yellow on the inside. There were eight stamens in the flower but only four made pollen. The remaining quartet were sterile and blackish nail heads known as staminodes. In the centre was an absurdly large stigma with four corners or lobes. I would learn that this gross and turgid organ gave the bush its species name, *Boronia megastigma* (biggest stigma). While the flower had an odd charm, what drew me to it for days was its strong smell, and I still can't describe it after almost half a century. It was so agreeable, so complex, and so enticing that I lost interest in mere roses and violets.

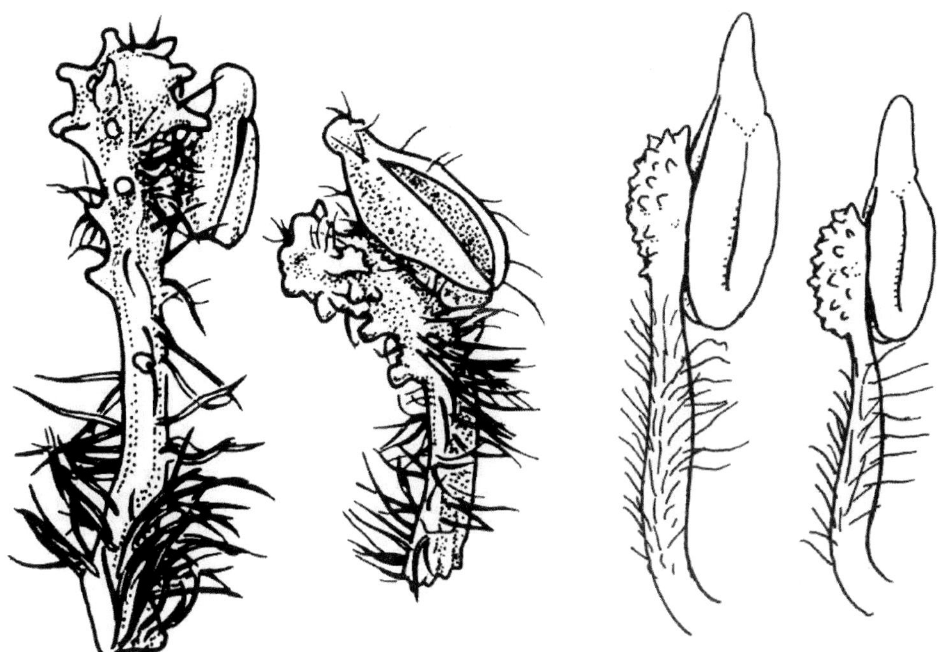

Drawings showing hairy and warty fertile stamens of different *Boronia* species. Reproduced with permission from: (left) drawing by Catherine Wardrop, *Flora of Australia* Volume 26, Australian Biological Resources Study, 2013; (right) drawing by A. Menadue, *Nuytsia* 12: 127, fig. 2, 1998.

Such was my introduction to the shrub most Australians knew as the brown boronia. The University's plant was not native to the Melbourne countryside. It had come a very long way, as its immediate ancestors were native to the Mediterranean climates of the swampier, southern woodlands of Western Australia, at the opposite southern end of the continent. As usual, the genus name, *Boronia*, tells you more about the loyalty 18th century botanists showed to each other than it does to the 148 species found only in Australia. Entomologist and botanist James Edward Smith (1759–1828) decided to honour Francesco Borone (1769–1794), a short-lived Italian plant collector who assisted in expeditions identifying the floras of Greece and Turkey.

At the time I went to Australia, boronias in general were among the few native species to evoke a sentimental response in the average citizen's breast. The most common species with pink and white flowers (*Boronia serrulata*, *B. pinnata*, *B. falcifolia*, *B. keysii*) were granted the title of native roses despite the early impatience and distaste British colonists often showed towards antipodal vegetation because it failed to resemble the lush and tamed countryside they'd left. In 1915, one councillor decided to name a Melbourne suburb, Boronia, after some plants on his property (11 species grow wild in the state of Victoria), and so it remains to this day. During World War I women enclosed dried sprigs of boronias in their letters to soldier boyfriends and husbands in Turkey or Europe. Had the plants been easier to grow they might have been planted to scent and brighten graveyards as first suggested by Ferdinand von Mueller (1825–1896), the first official botanist of, what was then, the colony of Victoria. In turn, von Mueller's name was honoured in 1924 with the species *Boronia muelleri*, native

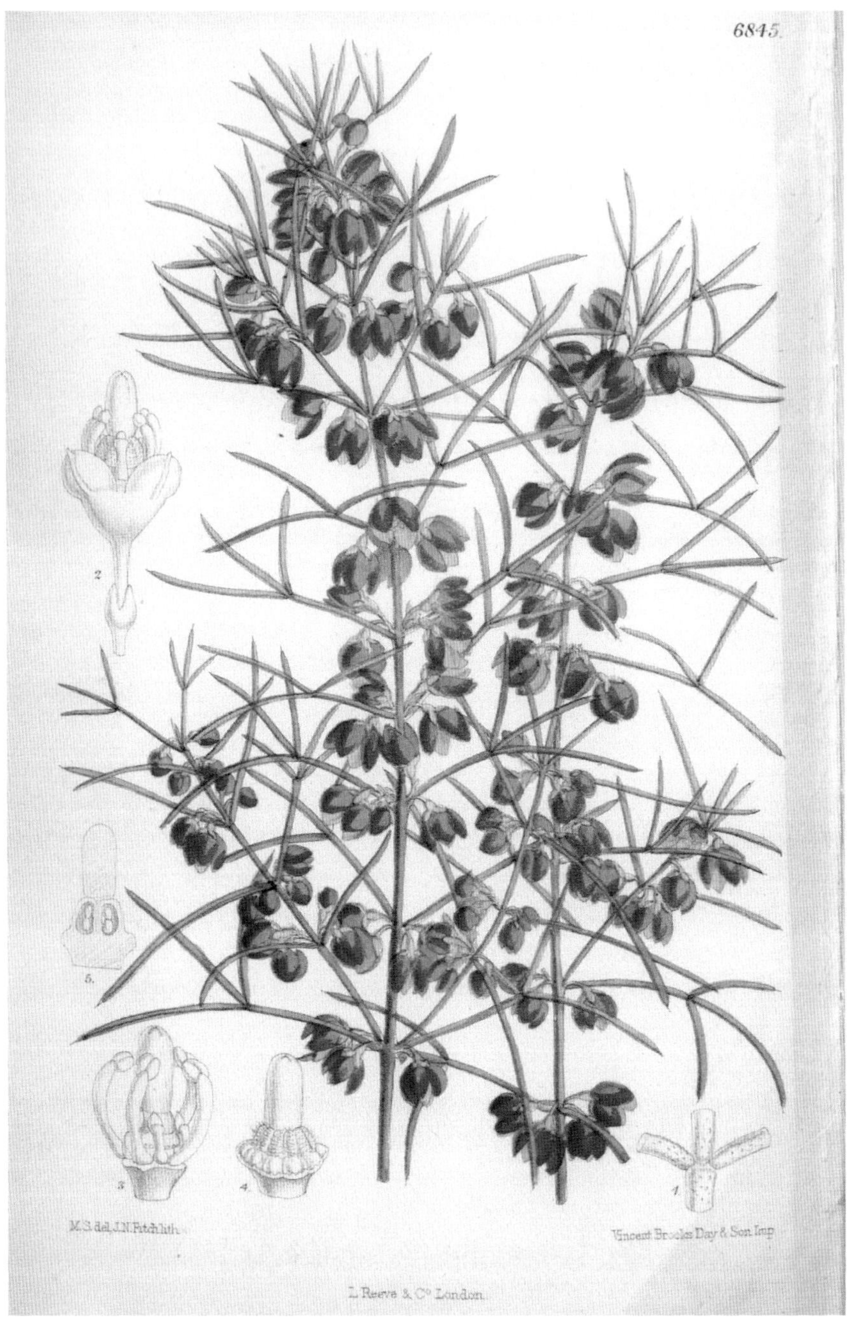


This classic plate of the red or Kalgan boronia (*B. heterophylla*) comes from a Curtis's Botanical Magazine (1885) and also shows 'half-flowers' (mid- and lower left-hand corner) depicting the central, exaggerated and swollen stigma surrounded by long, fertile stamens and shorter sterile stamens.

to the state of Victoria. The Australian children's author and artist May Gibbs (1877–1969) anthropomorphised the flowers in the book, *Boronia Babies*, in 1917. These boronia sprites ride on butterflies, use their four petals as parachutes and spread their perfume through the bush.

## THE BORONIA INDUSTRY AND ITS BIOCHEMISTRY

It's the brown boronia that people return to every time due to its distinctive fragrance. Von Mueller also thought its essential oils might be harvested and used to improve the aroma of black teas. While I was a graduate student in Melbourne, Western Australians could apply for licenses to pick sprays from bushes in their native habitat and it was a commercial success. Florists sold them for tussie-mussies and I understand they remain popular in Japan. More importantly, the flowering twigs went into jars of hexane solvent to extract the volatile molecules. A mixture of solvent and essential oils is called a concrete. Every time a laboratory separates plant fragrances from binding solvents the remaining liquid of concentrated and purified scent is sold as an absolute. In Australia, brown boronia absolute was combined with the scents of one of the sandalwoods (*Santalum*) and other components to make a perfume sold as Double L. My late mother liked it very much. I haven't seen Double L perfume in years. I understand the essential oils of brown boronia and sandalwood still accent new perfumes made by the Australian company Goldfield & Banks.

Boronia flower absolute remains an Australian export used in a range of globally marketed cosmetics. It also 'strengthens' essences derived from edible fruits to flavour soft drinks or tobacco. While flowering stems once sold for $10–15 a kilo, a 30 ml flask of brown boronia absolute may now sell for US$1,000. The era of licenced pickers in south-western Australia gave way to commercial plantations with much of the boronia industry transplanted to the cooler and moist island of Tasmania.

Beta-ionone is a prominent component in the scent of brown boronias and reminds people of bouquets of violets or strips of fresh cedar wood. In fact, a total of 160 different molecules have since been identified in these flowers under cultivation. This seems astonishing when we consider that the South African freesia (Chapter 1) offers a mere 50 compounds. When Dr Kaiser performed his analyses, over several decades, he usually used the flowers of only one or a few plants of the same species. Floral scent, like any other trait or character, are under the control of one to many genes in differently distributed populations belonging to the same species. Scent production in these different and isolated populations may vary in response to subtle and long-term changes in environmental pressures including climate, soil chemistry, predators and available pollinators. The inflated diversity of scent components in brown boronia must also be based in part, on what happens when members of different populations are grown, hybridised, and new breeds or cultivars are produced by and for the horticultural industry. Due to our domestication of dogs over thousands of years, we understand that man-made selection can alter frequencies in gene pools as effectively as natural selection.

Unfortunately, growing brown boronia, and the equally popular red or Kalgan boronia (*B. heterophylla*), is difficult as they succumb to a range of complaints when removed from their original wetlands in Western Australia, even when their descendants are transferred to well-watered Tasmania. The plant I harassed in 1977 died decades ago and was not replaced before I left Melbourne in 1985. When sold in Australian nurseries buyers are advised to treat boronias as short-lived shrubs.

What makes boronias such fragrance factories? Genealogy is a factor as they share a family tree with so many scented relatives. The flowers, foliage and fruit skins of some of the more than 2,000 species in the Rutaceae are among the most fragrant in the world. If you can hold a thin leaf or petal up to a lamp and see tiny holes transmitting the light you have found the many pellucid dots of the Rutaceae. They contain clear volatile oils. The glands that made these oils emptied them into tiny caverns or barrels, the pellucid dots. These barrels may release their contents into the air in hot dry weather or when a plant organ is bruised. They give us the familiar smells in the skins of citrus fruits as well as the scent of orange blossoms at weddings. As spices, Indian cooks use curry leaves (*Murraya koenigii*) and there is the characteristic zing of Sichuan peppers (*Zanthoxylum simulans*) in the cuisines of south-western China. Aegle is a name of at least six minor Greek deities. *Aegle marmelos* is the Bengal quince or bael, a fleshy fruit with an aromatic pulp. The 10 strongly odorous species of rue (*Ruta*) are native to Eurasia, northern Africa and Macronesia. They have an old history as flavourings, traditional medicines and insect repellents. Historically, rue's strong but natural smell invokes the supernatural. The common rue of the Mediterranean basin (*Ruta graveolens*) was once favoured in witches' charms, while the Catholic Church viewed it as the 'herb of grace' using fresh stems to sprinkle holy water. It is mentioned in the Talmud and it has appeared in folklore as a plant that protects from diseases and the evil eye.

Three scientists in Western Australia looked at the development of flower glands in six brown boronia bushes and the chemicals they made, publishing their results in 1995. They found pellucid dots inside most of the flower's organs, including the nectar gland. In addition, there were scent-making hairs and bumps (osmophores) on the surfaces of the petals and on the blackish heads of the sterile staminodes. If we liken scent molecules to notes of music then a boronia flower is like a well-rehearsed orchestra, as its different organs release different concentrations of scent chemicals at different times. The concentration of beta-ionone and heptadecene peaked as the flower opened, signalling that it was ready to receive pollinators.

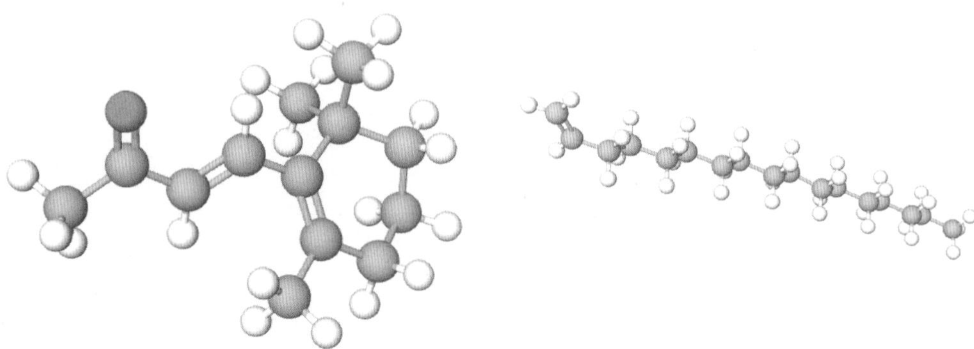

Beta-ionone (left) occurs naturally in the much-loved smell of brown boronia flowers, but so do another 159 scent molecules, including heptadecene (right). See how very different flowers use and modify the same molecular skeleton as Rupp's midge orchid secretes 8-heptadecene (see Chapter 4) which reminds some people of wax candles or aging fish in its pure state. Hydrogen (white), carbon (light grey), oxygen (dark grey).

The production of these same molecules collapsed as the withering flower shed its petals and male organs. Other pleasant compounds, like limonene, didn't increase until the ovary swelled into a ripening fruit.

There is plenty of room for more analyses of glands and chemicals made by the organs in boronia flowers. What about the dozens of species that are not of commercial importance? Their flowers usually lack four black staminodes. Instead, they have eight pollen-making stamens. Under magnification, these virile organs may look grotesque. resembling hairy, pimply goblins bent over the burdens of their pollen sacs.

## BORONIA AND ITS MOTHS

So where do moths fit into this smelly life history? Brown boronias devote so much of their resources to perfuming their flowers you'd expect that a shrub in bloom would be inundated with insects of all sorts. Surely, we should expect the bright butterflies of a May Gibbs illustration, commercial honeybees, or at least the pollen-eating hover flies to visit but those are rare occurrences. In the early 1980s I attended meetings of the Australasian Pollination Ecologist's Society. Greg Keighery, then a botanist for the Kings Park and Botanic Garden in Perth, was studying the pollination of many native plants. While he documented the insects and birds carrying the pollen of blossoms of most of the major groups of Rutaceae growing in Western Australia, boronias, so iconic to wildflower tours, left him perplexed. At one meeting, he confused us when he showed slides of tiny, whitish moths on the flowers of brown boronias. Greg described how these insects rammed the tips of their abdomens into the big, thick stigmas. He asked us, 'What's going on here?' No one replied, and I thought it was just another case of a codling moth laying her eggs in a flower that would never grow into a healthy fruit. Much later, in 1999, a paper comparing variation in oil components in brown boronias mentioned moths leaving pollen on the stigmas, but I wasn't paying attention at the time.

To be honest, I didn't think much about boronia reproduction for four decades until I saw the 2019 PhD thesis of Dr Elizabeth Milla in 2022. The second page offered a full-colour photo featuring a skinny little wisp of a moth inspecting the yellow interior of a flower of the brown boronia. White globs were visible on the maroon surface of the stigma. Those globs resembled the fluffy clump of pollen extruded by fertile anthers in the same flower.

I was looking at the photos of an adult shield-bearer moth in the family Heliozelidae. They are a family of day-flying moths, but they are not closely related to true butterflies. Molecular genetics suggests the ancestors of heliozelids diverged at least 80 million years ago. Their reasons for abandoning night flight remain a mystery. Perhaps they were always day moths, as their ancestors appeared long before the first bats (Chapter 10). Their caterpillars are leaf miners, each spending their youth chewing through the green layers inside succulent foliage. When they are ready to pupate they cut out a protective, shield-shaped portion of the leaf's blade and fall to the ground. In Australia, though, some heliozelid moths provide completely different nurseries for their larvae. Some females introduce their eggs into the stem tips of members of the heath family (Ericaceae) and the caterpillar turns the tender tissues into its own gall house after it hatches. A second group won't mature into winged adults unless they can invade the fruits made by members of the Rutaceae and they eat the developing seeds. Seed-eating caterpillars that attack fruits of Rutaceae are usually placed in the genera *Pseliastis* and *Prophylactis* and appear to be unique to Australia.

Dr Milla and her colleagues looked at 37 species of moths associated with the flowers of 15 *Boronia* species. Several *Prophylactis* species, associated with boronias, were unremarkable at first. They appeared to lay their eggs in developing fruits pollinated previously by other animals. Their caterpillars messed up fruit development like any parasite. Under magnification, though, females of the remaining moths proved to be so distinctive the team decided they were members of a group new to science and they deserved their own scientific names. What was so novel? Under a microscope, the abdomens of each female moth had its own, deep and longitudinal cleft packed with pollen.

Populations of flowering shrubs of brown, red (*Boronia heterophylla*) and yellow (B. *tetrandra*) boronias were targeted for close observations. That included making videos and sucking up more specimens with a pooter, a technique described in Chapter 4. During daylight hours, female moths visited the flowers of all three species. Sometimes the same flower hosted more than one moth, but their behaviour was predictable. There is a thin, low nectar gland ringing the base of the ovary. A moth may extend her hair-like proboscis to sup up a little of its fluid, but nourishment was never her primary goal. She was looking for a place to lay her eggs. This search could take her as long as 4 min in each flower. When some moths laid eggs they left a scar on the flower's skin. Something else was going on but it was harder to see. The cleft on her flexible abdomen was contacting the fertile anthers extruding puffs of pollen. Just before she left the flower, the mother moth crawled up onto the central, thick, lobed stigma and gave her own abdomen a squeeze. A glob of white pollen was pressed onto the stigma surface over half a minute. By the time the caterpillar hatched out weeks later the pistil's ovary was swelling and developing seeds. The larva burrowed into the fruit eating some seeds. Greg Keighery's weird observations had finally been vindicated decades later.

It's an intricate story. When most *Prophylactis* moths attack boronia fruits they are there only to exploit them for their offspring. In contrast, the new species of cleft moths arrive for the same reason, but they are the only insects now known to smear fresh pollen onto the thick and receptive stigma. Dr Milla watched infrequent visits by beetles, thrips, ants and flies, and they were all ineffective as pollen carriers. The butterflies depicted in the fantasies of May Gibbs never appeared. Yes, the caterpillars of the cleft moths must destroy seeds to live but they are not gluttonous and leave enough to produce the next generation of fragrant shrubs.

Relationships are expected to become even more complicated as research proceeds. Some *Boronia* species receive both pollinator and parasitic moths. The brown boronia appears to be very specialised and may have only one dependable species of moth pollinator. Some *Boronia* species may entertain two or more species of true pollinators. We do understand that relationships between a *Boronia* species and its pollinators depend on very localised ranges in Western Australia. While brown boronias are now propagated by cuttings and planted outside their natural distributions they won't set fruit in Tasmanian plantations as the shrub and its moth are not native to the island. The system appears to be so finely tuned to specific sites that a colleague living near Perth, Western Australia, reports that his brown boronias don't set fruit either even though his land grades into natural bush.

There's a very important part of this story no one can answer yet. We return again to Charles Darwin who is joined, this time, by decades of plant breeders whose experiments confirmed that 'Nature abhors perpetual self-fertilisation'. Inbreeding makes populations less able to respond to changes in their environment over several generations and may reduce the annual yield of healthy seeds needed to produce the next generation. The obvious question to ask is whether cleft moths are agents of cross-pollination? Are they programmed to gather pollen from a flower on one shrub and then extrude it onto the stigmas of blooms on a second plant? As they are such small insects, their flight patterns are almost impossible to follow. This question can be answered someday provided there is funding for a lot of long and expensive laboratory work analysing and comparing gene frequencies between parent plants and their seeds.

## ACTIVE VERSUS PASSIVE POLLINATION

The idea that an insect 'knows' it must transfer pollen to a stigma to produce a nursery for offspring it will never see is remarkable. It means there is a procedure in Nature called active pollination. The previous chapters in this book and Chapter 10 are all examples of passive pollination. Most animals visit flowers expecting only an immediate benefit. The reward is usually edible although the male gnats and wasps in previous chapters are lured by the false promise of sex. When animals, such as bees, collect food and drink for their offspring or siblings they take it home with them. They don't carry their larvae to open flowers and expect them to feed themselves. In all cases, the animal brushes against the pollen-laden anthers and bumps into the stigma as an indirect consequence of foraging or probing. The creature's contact with the stigma is not deliberate. It happens repeatedly because floral architecture gives the animal only limited space to manoeuvre.

In contrast, a cleft moth deliberately uses her abdomen to collect pollen then deposits grains on the stigma of the boronia pistil. Is this thoroughly active act of pollination unique to Australian boronias? No! It has evolved several times and may be far more common in Nature than we've assumed. If the molecular clocks proposed by geneticists are correct then fig trees (*Ficus*) and parasitic fig wasps (Agaonidae) have enjoyed a mutual interdependence since the Cretaceous. What we call a fig fruit is actually a chamber lined with male and female flowers. The 'belly button' at the bottom of a ripe fig was actually the natural portal (ostiole) through which female wasps entered and exited. The male wasps are wingless and once they mate with the female die inside the fruit chamber. It's the female wasp who gathers pollen from male fig flowers, leaves her birth fig and enters a second, immature fig (hopefully on a second tree) to spread the pollen on the female flowers and lay her eggs. She also dies inside the fig she pollinates. These insects are far smaller than any of the wasps described in previous chapters, so there's a good chance that if you've eaten a wild fig you've eaten a wasp cemetery. There are close to 900 species of figs in the world and their relationships with agaonid wasps must be successful whether the *Ficus* species spends its adult life as a huge tree, a woody vine or as a smaller shrub clutching the branches of another canopy giant.

## ACTIVE POLLINATION BY YUCCA MOTHS IN NORTH AMERICA

More importantly, cleft moths are not the only moths with careers in active pollination. Until these recent discoveries in Australia, the best examples came from North America based on research that began in the second half of the 19th century. *Yucca* species and their moths were literally the textbook case used to represent active pollination and obligate relationships between flowers and their pollinators. Botanist George Englemann (1809–1884) of the Missouri Botanical Garden and state entomologist Charles Valentine Riley (1843–1895) published their first observations of the yucca flower and females of *Tegeticula yuccasella* in 1872. With at least 50 *Yucca* species found only from North to Central America it was considered almost un-American for children of my generation to not know a version of the yucca-moth story as addressed in the natural history books of my boyhood.

Yuccas remain much appreciated in corners of the United States. The white, drooping and waxy blossoms of the soap tree (*Yucca elata*) are the state flower of New Mexico. As American gardeners grow more concerned about the cost of water to maintain lawns in an era of drought, cultivars of yuccas have become increasingly common and they are almost indestructible ornaments of suburbia. Most people grow the perennial species that look like sprawling rosettes made of sharpened swords and some naturalists continue to call them Spanish bayonets. However, they are not quite as American as apple pie. Yuccas are distant relatives of asparagus and their blanched petals are used in Mexican and Central American dishes. While I was a Peace Corps volunteer in El Salvador in the mid-1970s, I remember how the petals of *izote* (*Yucca gigantea*) were pickled with cabbage and carrots then served as a relish with cheese- or pork crackling-stuffed tortillas known as pupusas.

At first glance, the relationship between yucca flowers and moths in the family Prodoxidae looks almost identical to the boronias and their heliozelids. The yuccas can't make seeds without the right prodoxid moth, but not all moth visitors are pollinators. There are also 'bogus' yucca moths (*Prodoxus* species), first identified by Riley. Their caterpillars ruin yucca leaves or fruits. As usual, true pollinators in the genera *Tegeticula* and *Parategeticula* lay eggs in the flower before leaving pollen on the stigma. As usual, their caterpillars hatch inside the developing fruit by helping themselves to young seeds. As in the case of boronia pollination described above, some moth species are known to pollinate two to six different *Yucca* species, while others are extremely faithful and are associated with only one.

There are important differences, though. Unlike patient boronia caterpillars, yucca larvae hatch out within a few days and immediately begin eating immature seeds. Unlike boronia moths, both bogus and true yucca moths have expanded their natural distributions as people take the moths' host plants into parks and their private gardens.

Of even greater importance, unlike the cleft moths of boronias, true yucca moths do not feed as adults and they would *never* consider using their bums to pollinate a flower. The mouthparts of a yucca female don't suck. They are replaced by modified, tentacle-like organs she drags across an open anther to harvest pollen grains. The moth also secretes a liquid that lets her tentacles roll the pollen into a moist, easy-to-mould package containing up to 10,000 grains she stores in a hatch under her head. By the time she finishes harvesting pollen the package weighs

A prodoxid moth (*Tegeticula antithetica*) adds her pollen ball to the receptive stigma of a yucca (*Hesperoyucca whipplei*) pistil. Photo by Sherwin Carlquist. Reproduced under a Creative Commons License: https://creativecommons.org/licenses/by-sa/3.0/

10% of her bodyweight. Now she looks for an appropriate flower and uses her abdomen to bore a hole in the ovary laying more than one egg. Compared to those little boronia fruits, yucca capsules are massive, bigger than peanut pods, and can support far more caterpillars. Female yucca moths leave chemical messages (pheromones) in their flowers. An incoming moth must decipher the message and decide whether she will use a flower's ovary that has already been claimed by a rival. Once she's laid eggs a yucca moth ascends to the top of the pistil and chooses one of the three hollow stigmas. The tips of her tentacles scrape off some of the pollen package and she pushes the grains down the pistil's canal. Her pollen package contains so many grains she will continue to visit more yucca flowers on more plants, laying eggs and pollinating as she goes.

## ACTIVE POLLINATION BY GREYA MOTHS

As field research continues we may have to agree that active pollination by moths may be more common in North America compared to Australia. For example, *Greya* moths belong to the same family as yucca moths. The big difference is that greyas lack face tentacles and they only target North American plants in the rockfoil family (Saxifragaceae), quite unrelated to yuccas. These insects play an ongoing role in the laboratory of Professor John Thompson at the University of California at Santa Cruz. The interactions between *Greya* moths and the 12 species of woodland stars (*Lithophragma*) and 43 species of coral bells or alum roots (*Heuchera*) are far more variable than their yucca-visiting relatives. Furthermore, Thompson and his associates have graded each moth species on its relationships with different wildflowers. There are at least four categories. One moth species is no help to the flower and may even rate as an antagonist. Another goes through the motion of pollination but is rather inefficient. Then there are two moth species that rate from moderately to highly efficient as pollinators. The more efficient the moth the more exclusive it is to only one or a few wildflower species. One *Greya* species pollinates flowers while wearing grains on its head, but is not considered too important to woodland star reproduction. *Greya politella*, by comparison, is really committed to her role of plant breeder. Her abdomen ends in an elongated, egg-laying tail with a membrane sheath that carries the host flower's pollen. Yes, a form of bum pollination has evolved independently in a moth family in Australia and an entirely different family in North America. In North America, the pistils of different species of woodland stars vary in pistil length and architecture to accommodate the egg-laying equipment of different species of *Greya* moths.

While *Greya* moths are dependent on woodland stars to house and feed their offspring, some of these wildflowers seem 'disinterested' in these active pollinators. The dependence isn't always mutual and some wildflowers like the San Clemente (*Lithophragma maximum*), bulbous (*L. glabrum*) and slender (*L. tenellum*) woodland stars don't need moths at all to reproduce. Take the smallflower woodland star (*L. parviflorum*), for example. Like all woodland stars, it offers nectar but also accepts the visits of thirsty little bombyliids (bee flies) that accumulate the host flower's pollen passively while drinking nectar. Over 2 years of observation these long-legged flies made up 68–88% of all the insect visits to the flowers. In general, the flowers of the small star received so many visits from so many different insects that 77% of them developed seeds. With its specialised equipment and careful egg-laying behaviour, the *G. politella* moth often appeared to need the flower more than the flower needed the moth. Like boronia and yucca caterpillars, the larvae of *G. politella* weren't gluttons and ate only 15–27% of the seeds of its host.

The faithfulness of this moth to woodland stars, in general, was beyond reproach. Dr Thompson told me that they drank only the nectar of woodland stars and related coral bells (*Heuchera*). A female *G. politella* spent 10–30 min sipping from a single flower before turning around, poking her abdomen down the petal tube and laying her eggs. I suspect that, as she was so slow and precise, she couldn't match the repeated visits by the common and faster bee flies. A flower visited only by these flies was as likely to ripen into a seed-filled fruit as a flower containing a moth's egg. It looks like the specialised moth is only a necessity in some populations of woodland stars when bee flies and other visitors are rare or unavailable.

Thanks to Dr Thompson's own faithfulness to *Greya* moths we understand that the relationship between the insects and the flowers is scent driven, as in boronias and their cleft moths. *Greya* moths must locate their woodland stars by smell as such small wildflowers may hide among taller grasses or behind trees. Collecting and analysing the scents of all 12 known species of woodland stars in 94 populations identified 132 scent compounds. The number and volume of these compounds vary among species, but also within the same species from population to population. Ultimately, this may tell us how dependent some woodland stars are to different species of moths and/or whether the wildflowers are more likely to entice bee flies.

## WHAT ABOUT A MOTH AND A NIGHT-BLOOMING CACTUS?

We have begun to understand the business model between flowering plants and moths. The plant pays for pollen transport by offering 'childcare services' to the pollinator. Yes, it's a slightly destructive transaction but far more common in Nature than entomologists like Riley anticipated back in the 19th century. The evolution of this model becomes more comprehensible when we understand that these relationships are multi-textured, swinging from parasitism to mutualism, selecting entirely different lineages of moths and plants in two hemispheres.

Do you remember those awful pantry pests in the Pyralidae? The senita moth (*Upiga virescens*) is a pyralid that lives in the Sonoran Desert of northern Mexico and the American south-west, but she is no threat to corn meal or wheat flour. If you look at the tip of her abdomen, the scales are arranged like a feather duster. Yes, she uses it to collect pollen and rubs the grains on the stigmas of the senita or totem pole cactus (*Lophocereus schottii*), a favourite of hobbyists. This time we have a night-blooming cactus and the moth flies in ancestral darkness, like Blake's invisible worm. She is responsible for 75% of all pollinations with day-flying miner bees (Halictidae) picking up the slack if the flowers fail to shut up by dawn. The caterpillar spends about 6 days in the developing fruit, eating 30% of its seeds, and then does something unique compared to all the other moths we've seen. The caterpillar leaves the flower and drills into the cactus' succulent stem to sleep in her cocoon.

The study of the senita moth and its cactus wasn't published until 1998. The cactus has at least 10 more close relatives, so there may be more species of senita moths responsible for the next generation of totem poles. Mr. Blake, your invisible worms fly night and day among the flowers of native roses, yuccas, woodland stars and cactuses in both hemispheres. There are far more of them than you ever saw in your vision, but let's end the narrative with a chapter on hot-blooded pollinators that also prefer the night.

# 10

# When brutal bats bite big buds

*Hanging upside down like rows of disgusting old rags*
*And grinning in their sleep.*
*Bats!*
*In China the bat is symbol for happiness.*
*Not for me!*

– Bats, *DH Lawrence*

AFTER CELEBRATING MY BIRTHDAY AT A LOCAL RESTAURANT IN SAINT LOUIS one night, my wife and I returned home to find an invader. A small bat flew from room to room and had no trouble navigating two sets of stairs. We remained composed and sympathetic. I turned out the lights on the first floor and opened the front door. Within a few minutes, the bat's echolocation found the open space and exited our human cave forever.

Over the next week, I repeated the story far too often. The only enthusiastic audience of one was an editor of a botanical journal in Beijing. She believed that a birthday with a bat was a perfect combination and she told me how lucky I was. In Chinese, the words good fortune and bat are both pronounced 'fu'. It explains why stylised bats, often painted red (the traditional colour of prosperity, power, fertility and joy), embellished Chinese art for thousands of years. To enhance the symbolism further, a painting, scroll or fine piece of embroidered silk may combine bat motifs with ripe peaches, as the fruit represents a long and blissful life.

I'm sorry, Mr Lawrence, but you've been outvoted, as China is not the only place where traditional cultures treat bats as Nature's gifts of good fortune. The Pomo people of California said the needle-like teeth of bats could chew obsidian, turning shards of volcanic glass into the finest arrowheads. Other North American nations admired the swift acrobatics of their flight and told stories of how bats excelled at ball games played against other animals. In the southern Pacific, the large, flying foxes or fruit bats (Pteropodidae) are saviours in some of the myths of Polynesia. They put out fires, feed the starving, and rescue Samoan and Tongan princesses from cruel rulers. The Melanesian people of the Pacific also appreciate bats but let us leave that cultural reference for the chapter's end.

When you think about it, Western civilisation didn't show much affection for bats until midway through the 20th century. Scientists, conservationists and a few authors/illustrators of children's books attempted to improve bats' reputations as devourers of noxious insects and as owners of a sonar superpower much admired by the military. I would argue that these attempts were never as effective or long lasting as the good press given to large, predatory mammals like wolves, lions and tigers since the 1950s. Hollywood movies, comic books and Halloween decorations keep replaying creepy and threatening images. In particular, think of *Indiana Jones and the Temple of Doom* (1984). The hero and his team ride through a dry wilderness in India while the silly love interest points to 'flocks of birds'. Indiana Jones dismisses her nature appreciation with contempt. 'Those are giant vampire bats,' he warns. Watch the movie closely next time. Anyone with a little education in tropical wildlife can see the monsters are merely vegetarian flying foxes. In fact, only three out of more than 1,400 bats species on earth drink mammalian blood and they are not natives of Asia. They are tropical Americans distributed from parts of Mexico as far south as Uruguay. As a Peace Corps volunteer I lived in Central America for over two years and never had to remove a single vampire from my room or neck.

## BATS AS TROPICAL HORTICULTURISTS AND FORESTERS

To be fair, while it's unlikely many will ever change their opinions about the insects I've depicted in previous chapters, bats have gained a quantum of respect in the West thanks to the persistence of several conservation trusts, animal charities and educational organisations. In particular, Bat Conservation International, founded in 1982, speaks for endangered species and against human threats to remaining mega-populations. Their literature and website emphasise that different bat species perform many ecological services aside from consuming mosquitoes. Fruit-eating bats disperse seeds and should be considered planters of rainforests and cactus colonies. They are tree breeders as well. At least 530 bat species live partially or completely on a diet of pollen and nectar, pollinating the local flora including the wild Asian ancestors of modern domesticated bananas. In Mexico, long-nosed bats in the genus *Leptonycteris*, visit tiers of flowering branches of the blue agave (*Agave tequilana*) whose cultivated descendants give us mezcal and tequila. In south-eastern Asia, bat-pollinated fruit of the durian tree (*Durio zibethinus*) is an expensive and malodorous, but regional delicacy.

Scientific interest in bats as pollinators grew within the first half of the 20th century due, in large part, to observations by the great tropical botanist, EJH Corner (1906–1996). He was not content to catalogue and classify trees of the wet forests of the Malayan Archipelago. Corner's books and papers mined his field notes filled with observations of trees interacting with other plants, animals and fungi. In his classic work, *Wayside Trees of Malaya*, he introduces us to *Oroxylum indicum,* known commonly as the midnight horror tree, beka, the tree of Damocles, broken bones, Indian caper tree or scythe tree. It's a member of the catalpa/jacaranda family (Bignoniaceae). Here is what he has to say about the tree's short-lived flowers and its visitors:

> *'The flowers are nocturnal. The corolla begins to open about 10 p.m., when the tumid, wrinkled lips part and the harsh odour escapes from them. By midnight, the lurid mouth gapes widely and is filled with stink. Before sunrise, the corolla is detached and slips off over the long style. The flowers are pollinated by bats which are attracted by the smell and, holding to the fleshy corolla with the claws on their wings, thrust their noses into its throat: scratches, as of bats, can be seen on the fallen flowers next morning.'*

The description was so disturbing to the then young writer, Paul Theroux, that he incorporated these sentences into his short story *Dengue Fever*. Theroux forces the reader to wonder if a medicine made from the bark of this tree allowed a recovering invalid to see ghosts murdered during World War II. The flowers of the midnight horror sound unusually lurid but Corner described floral features we now expect to find in most plants pollinated exclusively by bats. These traits are shared by hundreds of unrelated plant species, whether they grow as trees, vines, branch-colonising epiphytes or as unusually tall herbs. Naturally, all are expected to produce flowers that open at night. When they have petals, the blossoms take the shapes of cylindrical tubes or nodding bells, but they almost always lack pretty patterns of pigmentation or bright colours. This is predictable as less than 1% of cells lining the inside of most bat eyes belong to the colour-detecting cones. To our eyes, most bat-pollinated flowers appear either white, or as a brownish-rust or in wishy-washy tints of green and yellow.

## FLOWER SCENTS AND FOOD REWARDS FOR BATS

Bat-pollinated flowers often smell bad to humans because their essential oils may mix sulphur compounds with derivatives of fatty acids and/or the sort of hydrocarbons our noses detect in rancid food. It's not surprising to learn that bat species living on diets of nectar and/or fruit usually have a much bigger olfactory bulb in their skulls compared to their insect-eating cousins.

As late as the first decade of the 21st century, many scientists assumed bat-pollinated flowers were pumping out molecules derived primarily from butyric acid, reproducing the nauseating smells one picks up in bad butter and sweat-laden gym clothes. In all fairness, let's consider the (atypical?) case of a passionfruit vine, the maracuja de restinga (*Passiflora mucronata*) of eastern Brazil. Most passionfruit species, including the domesticated vines we grow for pavlova garnish, bloom by day and are tinted in the brightest tones of blue, purple or scarlet but few people have commented on their scents. They attract bees or nectar-drinking birds for pollination. In contrast, the maracuja depends on Seba's short-tailed (*Carollia perspicillata*) and Pallas' short-tailed (*Glossophaga soricina*) bats for pollination yet people insist the flower smells more like raw pumpkin, fresh beans or even lemon cake. At least 39 different chemicals were detected in its night fragrance. Over a third of its scent came from the molecule ethyl linolenate, which is also found in several plant-based products. We humans find the smell of ethyl linoleate pleasant enough to include it in some commercial cosmetics. Tricosane also comprises almost 15% of this passionflower's perfume. The same chemical occurs naturally in plant essences we love including the therapeutic twigs of witch hazel (*Hamamelis*), spicy seeds of cardamom (*Elettaria*) and the flowers of linden (*Tilia*) when dried to make a comforting tea.

For any flower bat, a relatively simple diet of diluted sugars in nectar and protein and fat inside pollen grains has its benefits and drawbacks. On the positive side, the phyllostomid bats of the American tropics digest 99% of the nutrients in their nectar and more than 75% in their pollen. Yet there are physiological and physical challenges. These animals must contend with water regulation and electrolyte conservation producing a more dilute urine than do insect-eating bats. This requires remodelling the typical mammalian kidney. Then there's the problem of removing sweet droplets from the bases of floral tubes and bells. While most nectar bats

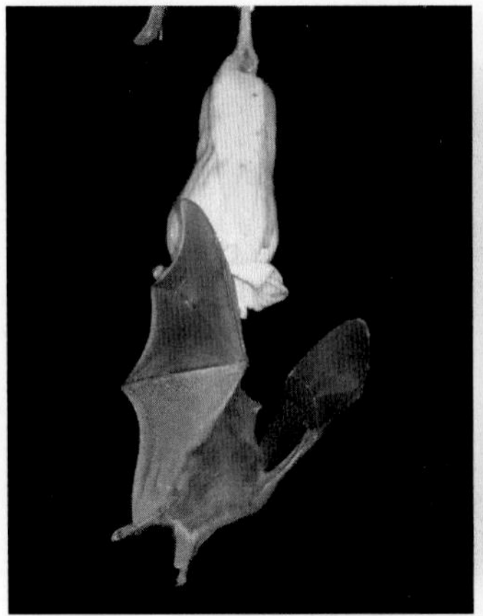

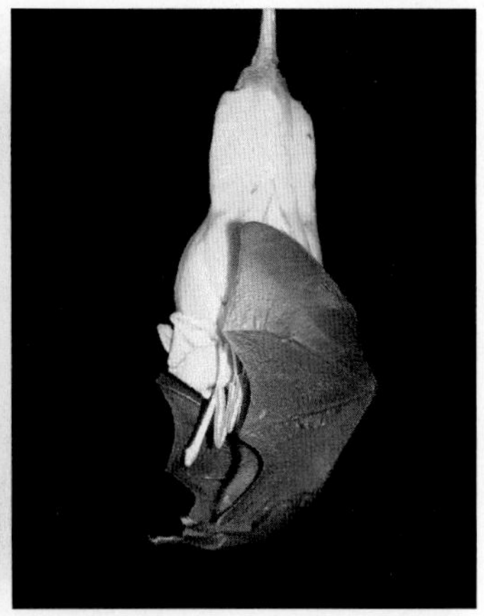

Sequence showing how a tailed tailless bat (*Anoura caudifer*) takes nectar from a South American nightshade flower in the genus *Trianaea.* A moment of clinging is required to reach the nectar at the base of the inverted bell. Photo by Nathan Muchhala.

have tongues tufted with thread-like brush cells that work together like a miniature sponge, the animal must be able to extend its tongue beyond the length of its mouth. The snouts of different species of phyllostomid bats sport slender tongues 5–7.5 cm in length. The tongue of the tube-lipped nectar bat (*Anoura fistulata*) from Ecuador is over 8.5 cm when unfurled and that is about the same length as the animal's body. Observers note that, regardless of location or species, nectar foraging is an unusually fast process with each bat spending no more than two seconds at each flower.

According to those committed to viewing them in the wild and analysing their droppings or fragments on their snouts, bat diets vary from species to species. Some insect eaters will take nectar or some fruit for the boost of energy afforded by simple sugars. Conversely, some nectar and fruit bats gobble the occasional moth, no doubt to supplement a diet low in salts and iron supplements.

Perhaps the most interesting case of an eclectic diet for any species occurs in the lesser short-tailed bat (*Mystacina tuberculata*) of New Zealand. They spend most nights foraging for flightless insects and grubs among the leaf litter of the southern cooler forests. However, they do have a taste for sweets and will drink the nectars of the native pohutukawa tree (*Metrosideros excelsa*) and the kaka beak (*Clianthus puniceus*, a member of the bean family (Fabaceae)), although these flowers are usually associated with native birds. More importantly, this bat is the only animal known to drink the nectar and carry the pollen of the dull-coloured flowering stems of the New Zealand wood rose (*Dactylanthus taylorii*). It's a leafless, root parasite that pops up through the litter on the forest floor. As if dusted with bleached flour, the short-tailed bat's muzzle becomes powdered with white pollen after a visit. It may be the plant's only surviving pollinator.

## GUARD YOUR NECTAR OR BE NOMADIC AND MIGRATE

In contrast, other bat species subsist only on floral foods and must contend with limited rewards. Few mass-flowering tropical trees remain in bloom all year. Moreover, as we approach the equator plentiful sunlight, stable warm temperatures, high humidity and generous rainfall encourages a greater diversity of woody species, but there is a limited amount of physical space for each tree to grow. The sheer number of trees belonging to the same species is low and its population is often spread over many hectares surrounded by unrelated competitors for elbow room and soil minerals. Nectar bats must search for their most preferred and productive flowers growing among so many other tree species that cater only to birds or insects, unsuitable and/or inaccessible to bats.

Furthermore, the number of nectar-feeding bat species in the same forest also increases as we approach the tropics as these animals also respond positively to year-round water and warmth. However, with a limited number of flowering trees available at any time, there will be competition for food resources. Competition for nectar and pollen occurs among bats belonging to the same and to different species. Some become rather aggressive. The black-bellied fruit bat (*Melonycteris melanops*), the Queensland blossom bat (*Syconycteris australis*), the long-tongued nectar bat (*Macroglossus minimus*) and several flying foxes (*Pteropus*) defend flowering trees or groves from competitors while flowering peaks.

Smaller, less aggressive species are nomadic foragers and must cover great distances at night, spending brief periods on widely dispersed nectar oases. As they appear to remember the best trees, at least while they are in bloom, ecologists refer to this behaviour as traplining. The trees in turn are often what plant ecologists call steady-state flowerers. While a single tree may bloom for weeks and have thousands of flower buds, it only opens a fraction each night and each flower is short-lived. The bat, in turn, remembers the rewarding trees and visits them for their limited but replaceable rewards each night like a trapper checking his line of snares. The combination of the bat's faithfulness to a particular tree combined with the tree's 'promise' of a small but consistent reward ultimately benefits isolated trees of the same species encouraging cross-pollination. A hot-blooded bat requiring sugar as a fuel is more likely to forage over greater distances compared to smaller, cold-blooded insects with bodies kept warm by the humid atmosphere. Some nectar bats take an extreme approach to seasonal rewards. Lesser long-nosed bats (*Leptonycteris yerbabuenae*) spend much of the year in tropical dry forests in Mexico before migrating as far north as parts of the Sonoran Desert in Arizona each spring to pollinate agave and cacti, including the iconic, giant saguaro (*Carnegia gigantea*). They make a round trip flight of more than 1,900 km.

## HEADS AND TAILS: NEW WORLD VERSUS OLD WORLD FLOWER BATS

When naturalists think of a bat adapted primarily to a diet of nectar, pollen and maybe some fruit, they expect its head will differ from that of an insect eater. As mentioned, a nectar bat has an extended tongue within its longer and narrower jaws and its larger olfactory bulb. We must also expect that a nectar bat must have larger eyes and fewer teeth than its bug-eating cousins. After all, the nectar bat does not need to snap up wriggling prey or crunch exoskeletons made of chitin.

Now, what about echolocation? Most of the flower bats in the flying fox family (Pteropodidae) in the Old World tropics tend to lack it. They communicate with squawks we can hear and their ears do not appear to be overly inflated compared to the size of their heads. Flying foxes are more likely to locate their food by sight and smell. In comparison, the long-nosed or leaf-nosed bats (Phyllostomidae) of the New World appear to retain some of the sonar of their ancestors. How do scientists interpret this difference? The evidence suggests that pteropodid nectar bats evolved directly from fruit-eating ancestors. In contrast, modern phyllostomids must descend from insect-grabbing ancestors. Of course, phyllostomids also find flowers using their senses of sight and smell. There is growing evidence that echolocation also plays an additional role allowing these mammals to differentiate between inedible leaves and fluid-filled flowers on the same tree. In fact, there is at least one tropical American plant that is so dependent on phyllostomids for pollination it announces its flowers with acoustical signals.

*Marcgravia evenia* is a tropical vine that twines around trees in the remaining forests of Cuba. A flowering stalk makes a dangling wagon wheel of 20 flowers and each flower has a dish-shaped leaf blade positioned just above it. These modified leaves, or bracts as botanists call them, produce strong, multidirectional echo signatures and are easily located by phyllostomid bats under captive experimental conditions. In the wild, Leach's long-tongued bat (*Monophyllus redmani*) has been photographed visiting the flowers of this vine and it may be its only pollinator in Cuba. It's unlikely that *Marcgravia evenia* is the only bat-pollinated plant in the tropics of the Western Hemisphere offering echo signatures to nectar bats, as there are another 59 *Marcgravia* species distributed from Cuba to Brazil. Leach's long-tongued bat isn't confined to Cuba either, and it frequents the forests of other islands in the Antilles. Where it is absent in the New World tropics there are plenty of related leaf-nosed species and I suspect some of them can also 'listen to the leaves' on other flowering vines and trees.

There's one more important distinction to make between the foraging antics of New World leaf-noses (phyllostomids) and Old World flying foxes (pteropodids). Leaf-noses appear to be the more gifted fliers and are more likely to hover as they drink nectar from a flower like a hummingbird. Flying foxes can't perform such aerodynamics because they have greatly reduced tails and that includes the loss of the uropatagium, the membrane of skin that links the two thighs of leaf-noses and most other bats giving them greater lift as they glide or flap. Flying foxes living on a diet of nectar, pollen and fruits are more likely to land and cling to twigs or whole flowers as they feed. This explains the scratch marks on the fallen petals of the midnight horror tree (*Oroxylum indicum*).

## THE KULUVA TREE BUDS BUT NEVER BLOOMS

Clinging hard to your drinking cup has its benefits. It may give you exclusive rights to a reward denied to night-flying insects according to a recently described interaction between pteropodid bats and a very special tree. This takes us to the islands of Fiji. Professor Sophie Petit of the University of South Australia and her team of four Fijian and Australian scientists found two Fijian islands offering the best sites to study the reproduction of the kuluva tree (*Dillenia biflora*). There are ~60 *Dillenia* species through the mainland of tropical eastern Asia and more islands in the South Pacific. Most produce large, flat, platter-like flowers with yellow or white petals. The little pistils mature into fleshy berries consumed by several animals of vastly different sizes

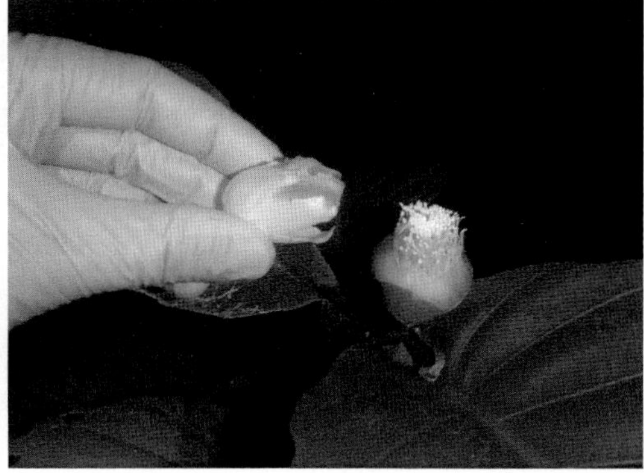

(Left) A mature flower of the kuluva tree (*Dillenia biflora*) whose petals never open. (Right) The dome of petals is removed by bats as the flower matures, exposing the many sexual organs inside the calyx cup. Photos by Sophie Petit.

and that may explain why some people call them elephant apples. Until recently, no *Dillenia* species was known to secrete nectar. A typical *Dillenia* flower is packed with 200–500 stamens attracting bees and hover flies with its generous pollen.

The kuluva tree, though, never opens its petals and each flower lasts only 24 h. A flower bud puffs up tall, sturdy and proud producing a white to yellowish dome of petals that stay locked in place due to the tight green sepals at their bases. While a few bud domes may add a splotch of pink pigment, floral development stops at this point. The photos I saw reminded me a little of a chef's hat. When flower buds refuse to open in nature botanists usually presume that the stubborn blossom pollinates itself in a self-incestuous but bisexual cross in which stamens curl over to rub their pollen onto the stigmas of the pistils. This is described as a cleistogamous (closed marriage) form of reproduction. Cleistogamy is often observed in some smaller plants including selected species of true violets (*Viola*), wild busy Lizzies (*Impatiens*), wood sorrels (*Oxalis*) and a few milkworts (*Polygala*). During the growing season each plant alternates floral production. Chasmogamous (open-marriage) flowers bloom first exposing themselves to pollinators encouraging cross-pollination with their usual range of colours, scent cues and edible rewards. Cleistogamous flowers tend to appear later in the season when overhanging, competitive vegetation begins to sprawl, making the habitat shadier, denser and far less appealing to sun-loving bees, birds and butterflies. Cleistogamous flowers resemble fat green buds that never open. In some violets, the closed-marriage flowers stand on such short stalks they simply flop onto the ground after ripening. The fruit splits open and the incestuous seeds spill under and around the parent plant.

Kuluva is no dainty little wildflower. Trees may grow more than 12 m in height and prefer sunnier sites at the very edges of wet forests and they eagerly colonise light gaps where old trees have fallen or been logged. Sophie and her crew noticed that the big buds produced a delicate perfume, unlike any known cleistogamous flower. In 2008, they paid attention to the words of a village elder on Waya Island who insisted the flowers were visited by bats during the summer months.

Sophie and her crew monitored 50 big buds that kept their petals intact and unmolested overnight. The usual experiments were made to determine the breeding system. If you slapped a bag over the flower so nothing could get in, they didn't set fruit (no closed marriage here). If you cut open the flower at night and self-pollinated it by hand, it would set fruit. However, they also produced berries if you brushed the pistils in one flower with pollen from a second tree so cross-pollination remained a distinct possibility.

To record what was going on at night the team placed four police body cameras equipped with infrared night vision and wide-angle lenses on the branches of nearby trees after sunset. Continuous filming was required. Videos showed invasive rats climbing up the branches, but they destroyed flowers. In 2017, a rare Fijian blossom bat (*Notopteris macdonaldi*) was filmed visiting a kuluva flower. It was identified easily as it's one of the few pteropodid species to have a long, skinny tail completely lacking a glider's membrane. The bat landed on sturdy clusters of kuluva leaves before crawling towards the flowers. The tape showed that the bat bit the closed petal dome then detached and flicked it off by throwing back its head. The stamens and pistils were now exposed. Over a period of more than four hours, more bats arrived biting 10 flowers on the same tree. The scientists examined the petal domes after they hit the ground. Each one had four or more puncture holes. Measurements showed they fit the width between the four canine teeth in a blossom bat's mouth.

What did the bats want from the flower? The scientists discovered that, unlike other species of elephant apples, kuluva flowers secrete nectar and quite a lot of it. Most flowers gushed around 500 μL of nectar. It may not sound like much but it's a generous puddle for a little bat. As the fluid was 10% sugar a bat had to drink it all to metabolise all the energy. Any remaining nectar left in these scalped and exposed flowers was consumed by late arriving moths and quarrelsome geckos. In other parts of their natural range, the buds of the kuluva trees were opened by insular flying foxes (*Pteropus tonganus*) or Samoan flying foxes (*P. samoensis*).

There may be at least two reasons why the kuluva presents its flowers as closed packages. While the appropriate bats are big enough, strong enough and probably smart enough to learn to open the flower, its petal dome is tough enough to shut out weaker nectar thieves like those moths and lizards. Sophie and her colleagues also suspect that a flower that makes its own lid is at an advantage on islands experiencing high rainfalls. The nectar sugar offered to the principal pollinator is already quite dilute and bats may lose interest if presented with an even weaker tipple while foraging on wet nights.

## PLEASE DON'T EAT BATS

The kuluva is one more tropical tree species growing in the centre of an ecological web. Field biologists continue to find that its nectar and fruit feed a circle of native animals. The same tree has a positive impact on the indigenous culture as it provides fast-growing timber and some traditional medicines. Conservationists think it may become an important tree to help restore rainforest degraded by farmland and teak wood plantations. Things should be looking up for its pollinators as well, but ... they are not. Yes, the traditional Melanesian people of Fiji appreciate their bats, but as a source of wild game cooked in coconut milk. I once read a story in an Australian newspaper that some of the Melanesian people of New Caledonia learned the culinary styles of their French colonists and have recipes for flying fox terrines. If the kuluva tree loses its pollinators it can't make enough seeds to recruit a new generation of saplings. Surely, that won't improve the quality of life for rural Fijians already dependent on a range of forest products.

Let's be candid, this isn't the only example of humanity shattering obligate interactions between species and it won't be the last. You can't blame these breaks exclusively on a taste for wild foods in the Pacific either. Visit the living rare plant collections and propagation houses in the National Tropical Botanical Garden on Kauai, Hawaii, and note the number of potted trees of native species that no longer reproduce in the wild. They must be hand pollinated, as so many species of indigenous bird-pollinators are extinct or reduced to tiny populations. While you can blame the original Polynesians for bringing rats and pigs to the Hawaiian Islands, and wiping out at least two species of native honeyeaters (Mohoidae) for feather cloaks, it's the European colonists who accidentally introduced mosquitoes carrying avian malaria and spread it further as nature lovers introduced and released exotic birds.

Can anything be done to achieve some balance between traditional appetites and environmental restoration? What if local law enforcement and education worked together to maintain a sustainable harvest of bats in the future? Other countries do much the same to protect their wild yet renewable resources of native timbers, mushrooms, game birds and freshwater fish. I'm not too optimistic right now. As gourmets on certain islands have decimated their native bats, they continue to purchase bodies from dealers operating in the Pacific who employ hunters or poachers. A dead flying fox costs US$40–50 these days. It's expensive but affordable for special occasions, just as our budgets make room for that holiday ham or turkey.

Yes, the book of Leviticus (11:13–19) classified bats as unclean fowl of the air, as are hawks, owls and hoopoes, never to be eaten but I doubt any rabbi has extensive influence in Fiji. The taboos of one culture can't be applied to a second to enforce conservation laws. Public health concerns might have some impact with government support. Consider how quickly the Chinese government shut down live game markets all over the country when bats, sold as exotic snacks, were one of the suspected reservoirs of the COVID-19 virus.

It is possible that a preferred food of pteropodid bats once posed a serious threat to the safety of the Chamorro people. They are of Micronesian descent and live on the island of Guam and the Mariana Islands. By the 1950s at least 60,000 Chamorros suffered from symptoms similar to Alzheimer's, Parkinson's and Lou Gehrig's disease (ALS). The condition was known as amyotrophic lateral sclerosis-parkinsonism dementia complex or ALS/PDC. It took up to 4 years before victims died and by then they could not swallow, breathe normally or speak. The condition has almost vanished but so has the native flying fox species on Guam, known as fanihi (*Pteropus mariannus*). These animals were hunted to near extinction.

That's not a smoking gun, of course, but there is a native plant, with a broad natural distribution through the Pacific, that has been implicated and the vegan fanihis disperse its seeds when they are not drinking the nectar of island trees. The queen sago or spiny leaved cycas (*Cycas micronesica*) resembles a palm, but it never makes flowers so it can't develop a true fruit wall around its seeds. It is a naked seed tree (gymnosperm) like the ginkgos and pines, mentioned in Chapter 1. When a bat carries off a seed it only gnaws on the juicy outer rind before dropping it somewhere to sprout into a new cycad.

Unlike most flowering plants, cycads have cyanobacteria living in their specialised roots. These microbes absorb nitrogen out of pockets of air in the soil and pass it on to the host tree as ammonia and nitrites. Nitrogen remains the essential element in the manufacture of enzymes and proteins in all organisms. In exchange, the cycad passes on sugars and organic carbon to nourish its bacteria partners. Cynobacteria have survived for hundreds of millions of years in all sorts of habitats by protecting themselves with a brew of toxins including Beta-N-methylamino-L-alanine, or BMAA. They also pass this molecule on to their host cycads and some of it accumulates in the seed's fleshy rind. When bats eat the rinds of cycad seeds, BMAA is stored in their fat cells. The Chamorro people once liked making bat's head soup and I remind you that up to 60% of a mammalian brain is made of fat. This is only a hypothesis, though, and several neuropathologists insist that the studies of Dr Paul Cox (Brain Chemistry Lab, Jackson, Wyoming) and the late, great chronicler of nerve disorders, Dr Oliver Sacks (1933–2015) are irrelevant. Dr Cox disagrees and thinks BMAA makes an excellent model to study the folding of bad proteins that may lead to conditions like Alzheimer's.

Should we be surprised if Cox and Sacks are proven correct? I don't think so. We now understand that some of the beautiful species of poison arrow frogs of the tropical Americas have toxic skins, in part, because they eat toxic insects that eat toxic plants. Of course, there are plenty of examples of animals thriving after ingesting the toxins in their food and I mention the next one because it is also a pollinator. Consider the fat bodies inside the abdomen of the lovely monarch or wanderer butterfly (*Danaeus plexippus*). As a caterpillar, it munches leaves in the hoya or milkweed family (Apocynaceae), so rich in the deadly class of steroids biochemists call cardiac glycosides. Arrow frogs and monarch butterflies display warning colours. The bats, alas, do not and one wonders if BMAA also harms them in the long run?

I'm only urging caution by issuing the usual reminder to all of us that our present and future welfare is often dictated by local natural history. In the past chapters, we have learned that interactions between flowers and some unusual animals, while wondrous, are never as rare or innocent as once seemed. Understanding flower–pollinator interactions may not seem like 'Big Science' but the act of pollination should remind us that all living things are connected. There are consequences when you ignore the little things that run the world.

# 11
# Conclusions

*Remember the most beautiful things in the world are the most useless; peacocks and lilies for instance.*

– The Stones of Venice, *John Ruskin*

A COUPLE OF DECADES AGO, THE MISSOURI BOTANICAL GARDEN INVITED ME to join a book signing session with a second author. A woman came to the table, picked up a copy of my book *The Rose's Kiss: A Natural History of Flowers* (1999) and started to flip through it. She stopped on page 143, glared at the illustration, put the book down hard and walked off with a look of revulsion. She had seen the pen and ink drawing of *Philoliche gulosa*, a South African horsefly with a tongue twice the length of its body.

Has anything really changed? I'd argue that her reaction, while exaggerated, still represents the most common response, and that includes from some people who also profess a deep love for Nature, their own gardens and the belief that they must donate to organisations committed to conserving biodiversity. Yes, tolerance of bats and a few insects has increased thanks to spectacular documentaries and social media posts that capture and magnify the intimate moments of life in the undergrowth, air and water. However, do you think these videos are responsible for winning over the majority or, as I suspect, do they appeal primarily to a loyal and consistent minority? Surely, I wasn't the only American kid predisposed to overturning rocks to see what crawled out long before I knew the BBC had a Natural History Unit.

Granted, a portion of humanity is drawn repeatedly to monsters. There has always been, and will always be, a place in Western art for monsters, even the smallest. However, it's my impression that that special place is still reserved primarily for purely imaginary beasts. Dragons are preferred to dragonflies, at least in China. Real insects and other small animals were incorporated into fine paintings during the great age of the Dutch still life (c.1600–1800). However, they were bit players used to accent flower arrangements, foods and crockery. Less well known is the great crossover between the natural history of little creatures and art made by painters like the German artist and entomologist Maria Sibylla Merian (1647–1717). She was known best for her detailed depictions of South American insects. Maria was clever and recognised the limited tolerance of her patrons. She made her larvae, beetles and cockroaches more acceptable by depicting them on or near sprays of splendid tropical blossoms. One art curator contacted me regarding a Merian retrospect in Saint Louis and was devastated to learn that those giant moths were not the pollinators of the intricate flowers of the soursops on the same sheet. While their caterpillars eat the tree's leaves, the winged adults do not feed at all.

## DISDAINING POLLINATORS WHILE LOVING FLOWERS A LITTLE TOO MUCH?

That's the irony. The flower stays elevated in art while its pollinators are less known. When poets take up the case of bugs and bats the results, as seen in this book's epigraphs, are amusing but rarely flattering or enlightening. As many of the flowers described in previous chapters appeal to our easily tweaked senses of sight and smell, they don't need to appear in high art to hold our attention. Instead, living flowers fit well into the commercialisation of crafts, hobbies and cottage industries. The range of our senses is limited, though, and so is our acceptance. If we could smell the little blunt-leaved orchid would we want them in our gardens to remind us of the last time we visited granny in the nursing home? Who would want a garden border that attracts mosquitoes? We learned to enjoy perfumes crafted from Australian boronias more than a century before discovering the identities and essential roles played by their moths. The mere sight of blooming bulbs, ice plants and daisies from South Africa increases their demand in our gardens and glasshouses. Would we welcome their beetles if they were introduced to our properties by accident? There is some room for new fashions ... up to a point. While the colours of fungus gnat-pollinated orchids are unexciting, their floral shapes are so unique that more hobbyists outside Australia and New Zealand are growing them in shallow pots or terraria, but I doubt the same people take pains to protect the invasive fungus gnats attracted to helmet orchids or the compost.

Sometimes that's all to the good. As we've noted some types of popularity contribute to pollinator decline. It's one thing to enjoy the taste of a bat-pollinated fruit in season. It's another thing to eat the bat especially when you lack the information to recognise all the free, ecological services provided by animals acting as unpaid plant breeders and forest planters. Of course, considering the ongoing loss of natural habitats and biodiversity in the West, it's hypocritical to critique the appetites of those living in more traditional cultures. Once again, we who live in the West must accept the bulk of the blame as we've wreaked destruction and depletion on other continents and islands for centuries. Classical historians agree that wild populations of silphium, a member of the dill and coriander family (Apiaceae) from Libya, was eaten into extinction by Greek and Roman gastronomes who rendered its roots for resins to spice up dishes. It's rumoured that the Emperor Nero was given the last known plant and the species or variety has been absent so long we still don't know whether it was one of the giant fennels (*Ferula*) or a non-toxic member of the genus *Thapsia*, often called deadly carrots.

Some examples are more recent reflecting our obsession to possess the strange and beautiful. While I hope you now admire the work of Charles Darwin and his rigorous descriptions of the biomechanics of orchid flowers, every tropical species he examined came from a wild source based ultimately on expeditions and commercial collections. Growing orchids or their hybrids from seeds following hand-pollination did not become a successful component of the horticultural industry until a few decades into the 20th century. I've written of the feverish 'orchidelirium' of the Victorians in my book *Wily Violets and Underground Orchids.* European nurseries sent out collectors for tree-dwelling epiphytes and this often meant cutting down the canopy and removing every orchid from the fallen limbs. One collector went as far as stealing wild cattleyas local people transplanted to adorn their churches.

The moral here is that any wild plant species can be loved to death. Mere appreciation of the beauty of an endangered flower won't save a rare species, or any of the other organisms it needs to complete its lifecycle, especially if plants are not allowed to grow and vary where they are most likely to produce new generations. The beauty of evolution works best *in situ*, less so in pots.

## VARIATION AND NOVELTY

It is tempting to believe that we live in a world of *the* sun moth, *the* boronia, *the* fungus gnat, *the* greenhood orchid and so forth. If you do, you are ignoring the fact that there are hundreds of species of sun moths, boronias, fungus gnats and greenhoods within Australasia. What happens when we compare the variation of novelty within members of the same lineage?

The interrelationship of the blunt-leaved orchid and its mosquito species looks unique until you compare its flower shapes, nectar spur lengths, petal colours and scents to the other *Platanthera* species in North America with bee, moth and/or butterfly pollinators. If the quantity and biochemistry of flower shapes, odour particles and pigment chemicals are so easy for human plant breeders to shift after a few centuries of tasteful breeding, surely the selective foraging of native pollinators can have the same impact on wildflowers over far longer time periods. Indeed, some wildflowers in South Africa appear to be doing this in the short term under natural selection. The same species of sundew or native daisy varies its frequencies of petal colours across its natural distribution because different species of beetle pollinators have more limited ranges and these insects have their own favourite petal colours.

In the long term, we learned in Chapter 1 that evidence provided by amber fossils from Cretaceous deposits, showing the earliest variations among flowers and flower-visiting insects, goes back more than 100 million years. Organic evolution across vast timescales brings the possibility that different types of selection might shift gene frequencies in a population, and that includes the genes controlling some of the earliest phases of development occurring in shoots making flower buds and during stages of insect metamorphoses. Contemplate the much-exaggerated lengths and forms that flowers and animal bodies take. We call this exaggerated development *hypermorphosis*. It's the same process that gives us the lady slipper's overblown sac or the elongated tongue of the little tube-lipped bat. Or consider the developmental process of repeating the same body segments or fibres over and over, making structures longer, broader or more numerous. That process is called *metamerism*. Consider the scattering of so many nectarioles on petal skins or the multiplication of big nectary knobs forming a crowded circle in a flower of Brown's peony. Then there are obvious examples of certain organs going absent or appearing much reduced. Think of the fewer and shorter teeth in the skull of a leaf-nosed bat and note how even reduced organs are recycled. Remember the two, tiny secreting lobes in a midge orchid flower? They are the remains of sterile stamens and now they offer cocktails to chloropid midges.

Yes, we now know not to regard Australian organisms as 'bizarre' or more extreme in form or behaviour, compared to the biota on other continents. On the other hand, should we expect something atypical concerning flower–pollinator interactions in Australia compared to the rest of the world? I hope not.

## HOW UNIQUE ARE AUSTRALIA'S POLLINATION STORIES?

Let me offer my thoughts. First, I think we should view all pollination systems, in general, as understudied. Some pollination systems seem strange to us only because we don't know how extensive they might be unless we focus on the right location and that includes the right season. Consider those young wasp queens visiting Brown's peony each spring in Oregon versus the ageing worker wasps visiting felwort flowers during a Yunnan autumn versus male wasps attempting to copulate with the orchids of southern Australia as spring grades into summer.

Second, in previous chapters I've hinted how many unrelated plant species may depend on guilds representing groups of pollinators with similar needs in different parts of the world. Therefore, stereotyping Australia as the centre of weird interactions between plants and animals is silly when you consider how similar pollination systems crop up in isolated parts of the globe. Woodland shade, moist soils, mushroom growth and low temperatures bring out the fungus gnat-pollinated flora in Australia and Japan. There are shrubs and wildflowers in South Africa and Australia pollinated by wingless mammals. Then there are those equally extraordinary trends in pollination lacking obvious doppelgangers in Australasia. Consider the hundreds of species of painted wildflowers in South Africa competing for the exclusive attentions of dozens of species of monkey beetles.

Once you incorporate lifecycles and natural history into a broader global view, doesn't it all becomes relative? I think so. Can you really say that the relationship between Australian boronias and sun moths are more bizarre than North America's totem pole cactuses and their senita moths? Which is weirder, the Australian spider caladenias luring male wasps with the false promise of sex or the flowers of a Chinese dendrobium inviting female wasps by copying the smell of a tasty but nervous bee on the defence?

## TOWARDS POLLINATION NETWORKS OR LANDSCAPES?

A younger generation of ecologists now speak of pollination or pollinator networks. Once you are able to identify the animal and flower species interacting with each other, or find someone willing do it for you, it becomes a matter of observing, collecting and analysing who visits what and which ones leave pollen behind on a pistil's stigma. Results are complex when you monitor many insect species sharing the same season and site with a dozen or more plant species. Different plants with overlapping flowering periods often share or compete for the same pollinators. To make a diagram explaining degrees of interdependence you draw lines to connect the pollinators to each plant they visited (see the figure below). However, you must also indicate which plants prove to be the most popular and which animals are the most commonly observed visitors. The resulting and unequal web of interactions may vary annually when

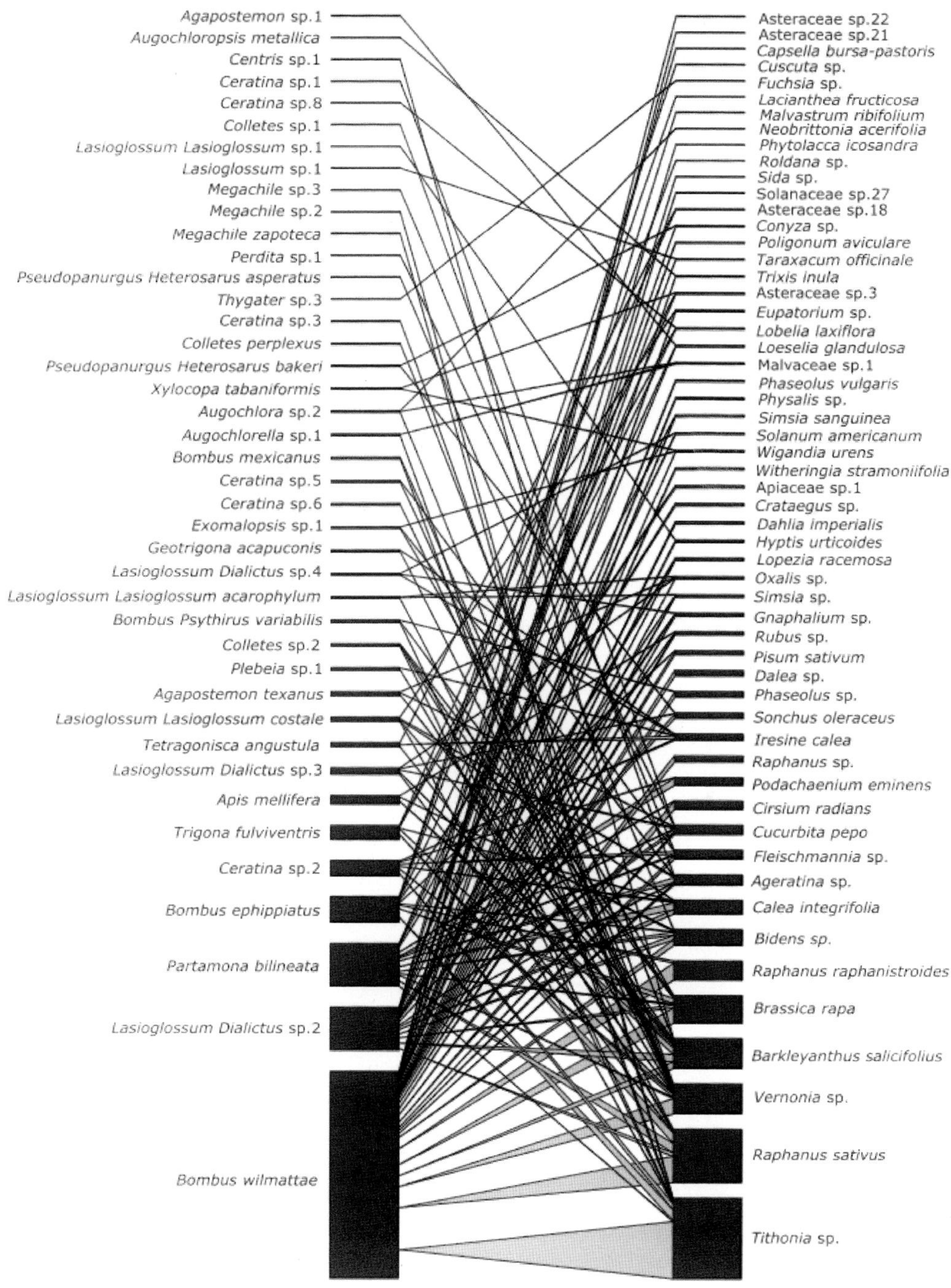


Pollination network from the Guatemalan highlands with bee species on the left and crop and wildflowers on the right. The thickness of bars and connecting lines indicates which bees are the most common visitors and which flowers are visited most often. Provided by Patricia Landaverde-González and Natalia Escobedo-Kenefic.

surveys go on for years. It works for fields of domesticated crops adjacent to native vegetation too. The finished diagram of interconnected relationships is your pollination network.

In contrast, most of my work over 47 years examines one flowering plant species and its pollinator(s) at a time. I'm happy to work with this new breed of researchers when they invite me, but I won't apologise for being a bit of a plodder. I think we may be on the verge of something new and better thanks to the last 150 years of *accumulated* plodding. Perhaps the time has come for us to start thinking about 'pollination landscapes' as circumscribed by geography, topography, climate and season. For example, we can say with some confidence that while the American tropics feature many pollination landscapes, those systems incorporating leaf-nosed bats are major contributors to floral diversity. Likewise, the winter woodlands of eastern Australia feature a landscape of fungus gnats and their orchids. Temperate bush from Australia's eastern to western coasts are more likely to show a landscape of male wasps and their orchids from spring to early summer.

## LOOKING FOR POLLINATION IN ALL THE RIGHT PLACES

In any case, why don't you consider joining us as an enthusiast or even a citizen scientist? Past generations of naturalists in both hemispheres experienced discovery and made an impact. You won't need your own ivory tower but you might want some training first. There are universities, museums and botanical gardens offering courses in the basics of plant and animal identification. A few offer classes in pollination biology, but you may be surprised at the high quality and clear language of recent books, not to mention the array of online tools and apps for identifying and logging nature (for which some of the images in this book were originally tasked). Some resources take a general approach. Others let you focus on orchids or on plants preferred by bees, birds or moths.

If you'd rather work alone and follow journal publications, you're also welcome. Basic equipment and chemicals aren't that expensive, and any room with decent ventilation, a sink, shelves and a couple of tables can become a laboratory. Funding and learning to use the new cameras and microscopes is, of course, a challenge.

There are extra benefits to consider. By knowing the lives of some of your non-human neighbours, and sharing your data with like-minded associates and administrators, you might be helping to protect that refuge or reserve down the street from future development. It's always exciting to be a co-discoverer of new species, but I think it's even better to illuminate a critical aspect of their lifecycles. Spying on the sex lives of fellow humans is reprehensible but the work we do for biodiversity has social value and may make you a raconteur of interest at parties. Always arrive with photos on your phones. When people ask me what I'm doing in retirement I tell them I'm writing pornography.

'Does your wife know,' they ask?

'Sure,' I reply, 'I let her read each chapter.'

They always look befuddled. 'Um, oh, and what does she say?'

I reply, 'She thinks it's pretty good but loses interest after the wasp flies off with the pollen.'

# Glossary

**Absolute.** Purified concentrated scent (essential oils) with all water, wax or cold fat removed.

**Adnation (Adnate).** When two different organs in a flower fuse together during the earliest stages of bud development and they stay interconnected after the flower opens (e.g. a petal fuses to a stamen).

**Amber.** Prized gemstones made of fossilised tree resins that may contain the remains and casts of plants, animals and/or fungi. The oldest amber is Cretaceous in origin and associated with the once dripping resins of extinct cone-bearing trees. The youngest is Tertiary and often associated with the resins of both cone-bearing and flowering plants.

**Androecium.** The ring or whorl of pollen-making stamens in a flower.

**Anther.** The lobed box, usually found at the tip of a stamen, that manufactures and protects pollen grains.

**Anthocyanins.** A subclass of flavonoid pigments composed of water-soluble molecules colouring some flowers, fruits and leaves red, pink, blue or purple.

**Betalain.** One of four major groups or classes of plant pigments. We see betalains as pink, red and purple petals. They are made almost exclusively by plants in the order Caryophyllales including sundews, cacti, portulacas, carnations, bougainvilleas and carpet weeds. Betalains also make the deep burgundy one sees in slices of beetroot. In flowers in other families, the same colours are more likely to be made by a class of molecules known as the anthocyanins.

**Boss.** A smooth, raised bump or hump on the upper surface of the lip petal of some orchids, especially helmet orchids in the genus *Corybas*.

**Brachycera.** An ancient suborder within the order Diptera containing ancestors of most of the fly families that visit flowers as winged adults (e.g. hover flies, hunchback flies, tangle vein flies).

**Bract.** A highly modified leaf that forms under one or more flowers or flower stalks. They also form protective covers over the ovules of gymnosperms.

**Calyx.** The outermost whorl of most flowers, often resembling a cup or star of green leaves once the flower is open. The calyx is made of segments called sepals forming the protective bud case of most flowers.

**Carpel.** One or more female organs usually located in the centre of every bisexual (perfect) or female flower. Each carpel has a receptive tip for pollen (stigma) and a chamber for seeds (ovary). In most flowers, the stigma interconnects to the ovary by way of a neck known as a style. When two or more carpels fuse together (connation) in the same flower the compound structure is called a pistil.

**Caudicle.** Usually one of two stalks made inside an anther of an orchid flower. Each caudicle connects the pollinia to a sticky viscidium plug.

**Chloropidae (chloropid).** A family of eye gnats or eye flies found throughout the world in the order of true flies (Diptera). Adult females of many species drink the tears of terrestrial animals, the fluids of weeping sores and/or the nectar of flowers.

**Cleistogamous.** Flowers that must always self-pollinate because their sepals and petals never open. They look like big, fat, perpetual buds until they fruit.

**Coalescence/Connation (Connate).** When two or more organs in the same floral whorl or ring fuse together and remain together once the bud opens. Connate whorls of petals or sepals most often resemble tubes, funnels or bells, while connate carpels form compound pistils.

**Column.** A floral organ made by fusing together at least one fertile stamen to the neck (style) of a carpel or pistil. In most orchids, the column is made by fusing together one pistil, one or two fertile stamens and either one sterile stamen (staminodium) or two sterile stamens known as staminodes.

**Complete flower.** A flower containing four kinds of organs (sepals, petals, stamens, carpels) arranged in discrete rings (whorls) or as a continuous spiral.

**Compression fossil.** Typical fossil found in slabs of sedimentary rock that underwent compression, usually over periods of millions of years, with the long squashing process often leading to distortion of the organism's original anatomy.

**Concrete.** The combination of scent molecules and the solvents or fats used to extract them. Historically, the process once meant allowing flowers on screens to dry and release the scents over a bucket of lard. In modern steam distillation the scent will be diluted with a certain amount of water.

**Connation** *see* coalescence.

**Corolla.** One or more rings or whorls of petals found in between the androecium and calyx in perfect flowers.

**Cretaceous.** The last period of the Mesozoic era beginning about 145 million years ago between the Jurassic period and ending with the arrival of the Cenozoic era about 66 million years ago. Most palaeontologists agree that the Cretaceous gave us the earliest fossils of flowers, the last big dinosaurs and winged pterosaurs.

**Cypripedioideae.** A subfamily of orchids in which all the flowers have two, pollen-making anthers and one large, sculpted staminodium located in between the two fertile anthers. The lip petal appears inflated resembling a slipper or shoe.

**Diapriidae (diapriid).** A family of tiny parasitoid wasps in the bee, ant and wasp order Hymenoptera. The pregnant female lays her eggs in the larvae of other insects, especially the maggots of fungus gnats.

**Elaiophore.** A gland that secretes oil droplets instead of nectar.

**Empididae (empid).** A family in the order of true flies (Diptera) known as dagger, balloon or dance flies. The male usually courts females offering them edible prey or a balloon of silk.

**Essential oils, ethereal oils or volatile oils.** Scented extracts that are most often derived from seeds, flowers, fruits, leaves, bark and wood resins. Each extract consists of one to many different kinds of lightweight molecules that volatilise or evaporate rapidly at room temperature. The majority are not true oils but form homogenous mixtures when added to liquid oils and cold fats.

**Exine.** The outer wall of most pollen grains made primarily of rods of sporopollenin molecules. It covers the intine wall.

**Filament.** A stamen's sterile stalk that usually bears the fertile anther at its apex.

**Fitness (reproductive fitness).** The outcome between genetic variation and natural selection in which organisms leave the most fertile offspring over successive generations.

**Flavonoids.** Plant-made chemicals each based on 15 carbon atoms. Some make pigments we can't see as they reflect the ultraviolet range, but one group called the anthocyanins gives us all the visible flower colours except for true yellows and oranges.

**Fungus gnats.** Members of several true fly families (Keroplatidae, Sciaridae, Mycetophilidae) that lay eggs on mushrooms or in humus where fungal threads (hypha) form tissue webs and skeins known as mycelia. There are also species of fruit or vinegar flies (Drosophilidae) that are dependent on fungi for maggot nurseries. These insects are often responsible for the pollination of plants that bloom in deep shade.

**Gynoecium.** The central whorl of carpels in a flower. Individual carpels may fuse together to form one compound pistil.

**Gymnosperm.** Non-flowering seed and pollen-making trees and shrubs with ovules never inside ovaries. Their ovules form on individual scales and may dangle naked from stalks (ginkgos) or are protected and partially concealed by spirally arranged but sterile bracts forming cones (pines, cycads, cedars).

**Guild.** Communities of different species living together in the same place and exploiting the same limited resource. They may share and/or compete for the same resource. For example, a mountain on which many unrelated wildflower species are pollinated primarily by bumblebees. Regardless of species, the plants bloom only when worker and/or queen bumblebees are active, attracting them with similar colour patterns and scents while offering edible pollen and/or nectar.

**Heliozelidae.** A family of small, day-flying moths that includes the cleft moths. Female cleft moths pollinate *Boronia* flowers leaving eggs on the flower's ovaries. The caterpillars eat some of the developing seeds.

**Hypermorphosis.** A form of evolutionary development encouraging a period of extended growth and development producing a much-exaggerated shape or length at maturity. For example, the long brush-tipped tongue of a flower bat or the much-elongated sepals of a spider orchid (*Caladenia*).

**Hypha (plural hyphae).** Thread-like cells or chains of cells of fungi emerging from spores. When hyphae grow together and form a tissue that performs the same function (reproductive, feeding, pathogenic etc.) it is known as a mycelium. When hyphae unite with each other and combine their nuclei briefly they develop into fruiting bodies, such as truffles, mushrooms and toadstools.

**Imperfect flower.** A unisexual flower, whichs makes either fertile stamens or an ovule containing carpels but never both at the same time.

**Incomplete flower.** A flower that makes less than four kinds of organs. For example, many wind-pollinated flowers do not make petals. Incomplete flowers can also be imperfect flowers if they don't make either stamens or carpels.

**Intine.** The inner wall of most pollen grains made primarily of cellulose and its derivatives.

**Krateridae.** A family of flower-pollinating beetles with a fossil history extending back to the mid-Cretaceous (98 million years ago).

**Lip petal or labellum.** The third petal in an orchid that does not resemble the two lateral petals and is always located opposite the pollen receptive surface of the flower's column. In slipper orchids, the labellum forms a shoe-like chamber while in most other species the lip petal is a landing platform that attracts pollinators.

**Metamerism (metameric segmentation).** A form of evolutionary development resulting in a 'copy, paste, and repeat' pattern of the same body part or unit. It explains, for example, why millipedes have so many jointed body segments each with two pairs of legs.

**Morphotype.** A group of individuals sharing the same physical/anatomical features. As there are so many different insects, and so few entomologists to describe and name them, the first step in understanding an understudied group of obviously related specimens is to find traits or characters they share and segregate them into discrete morphotypes for future classification.

**Mycelium (plural mycelia).** In fungi, when hyphae form a tissue with specialised functions. Different mycelia may unite to make multi-layered reproductive organs like mushrooms or truffles.

**Myrtaceae.** A family of woody plants including the classical myrtle bush of the Mediterranean basin. These plants show their highest diversity in Australia with hundreds of species of eucalypts, tea-trees, broom honey myrtles, bottlebrushes, and many other shrubs and trees.

**Natural selection.** Regional fluctuations (climate, shade levels, disease, predators etc.) in an environment over time that ultimately determine which members of a population are most likely to live long enough to reproduce and pass on their genetically determined traits to the next. Pollinators can represent a form of natural selection as their foraging behaviours and preferences ultimately determine which flowers are cross-pollinated inheriting genetic variability. Flower shapes, colours, scents and blooming periods may also represent the results of natural selection as they ultimately determine which pollinators or their offspring are fed enough to pass on their traits to subsequent generations.

**Nectar.** Nutrients diluted with water secreted by a nectary gland. They are usually dominated by sugars with some amino acids, vitamins and minerals. Regarded as a 'reward' for visiting animals that then carry the plant's pollen.

**Nectariole.** Diffuse skin cells or cell clusters, often surrounding or lining microscopic pores. They are scattered over the surfaces of petal skins and other flower organs releasing microscopic droplets of nectar.

**Nectary.** Discrete and specialised flower glands that secrete nectar. A nectary may consist of more than one layer of plant tissue, but it usually links to a vein or strand of sugar-conducting tissue known as phloem. Unlike nectarioles, the location of a nectary is fixed and genetically prescribed, and is always found in the same location.

**New Jersey Light Trap.** A device invented in 1927 by the New Jersey Agricultural Station to trap tiny insects (mostly mosquitoes) from dusk until dawn. The trap may be hung or mounted on poles. The insects are attracted to the light and sucked in by a fan (added in 1930) depositing them in a removable jar at its base.

**Orchidaceae.** The largest or second largest family of flowering plants with over 20,000 species. None have woody tissues. Most produce bilaterally symmetrical flowers with one exaggerated lip petal. The male and female organs fuse together (adnation) as a column that both releases and receives pollinia. The family is a source of many ornamental hybrids.

**Osmophore.** Whole organs or modified clusters of skin cells on the surface of floral organs that release characteristic scent molecules and are often visible to the human eye.

**Ovary.** The swollen base of a single carpel or compound pistil housing one or more developing seeds within one or more chambers. Most ovaries have one or more styles that interconnect the ovary to the pollen-receptive stigmas.

**Ovule.** An unfertilised seed attached directly or indirectly to the carpel wall in flowering plants. Each ovule contains at least one embryo sac. Each ovule has a pore (micropyle) at maturity allowing sperm-containing pollen tubes to enter and fertilise cells in the embryo sac.

**Pellucid gland or dots.** Shallow barrels made of cells that secrete aromatic oil-like substances embedded in foliage, flowers and fruit skins. They appear as translucent dots when a leaf is held up to the light and are especially common in the citrus or boronia family (Rutaceae).

**Perfect flower.** A flower that has at least four whorls or rings of each organ: sepals, petals, stamens and carpels.

**Petal.** The sterile organs in a flower forming one or more rings (corolla) sandwiched between the calyx (sepals) and the stamens (androecium). They are often, but not always, the largest and most colourful organs in a flower.

**Phloem.** Tissue in which cells form chains inside the stems and roots of ferns, and almost all seed plants. Phloem cells usually transport nutrients made in leaves down to underground roots. They also appear to move sugars into nectar glands.

**Phyllostomidae (phyllostomid).** A large family of tube-, leaf- and long-nosed bats of the Western Hemisphere. Different species have extremely different diets including fish, frogs, blood, fruit or flower nectar and pollen. Those that visit flowers appear to have retained some echolocation.

**Pistil.** A compound organ produced by the fusion (connation) of two or more carpels. The fused carpels share a common ovary.

**Pollen.** Grains made inside anthers. Each grain contains two or three cells and one or two of these cells are sperm. Most pollen grains are protected by a double wall. The outside wall (exine) is made of sporopollenin and the inside wall (intine) is made of cellulose products.

**Pollinarium.** Compound pollen dispersal unit in most orchids and some members of the oleander/milkweed family (Apocynaceae). It usually consists of one or more pollen masses (pollinia) and a sticky plug (viscidium or corpusculum). Many flowers that have pollinaria connect their pollinia to the viscidium with sterile stalks. Connecting stalks in orchids are either made by the anther (caudicles) or by the stigma lobe (stipe). In unrelated milkweeds, the stalks are known as translator arms.

**Pollination.** A frequent and entirely natural act in which a pollen grain (containing one or two sperm cells) is deposited on the receptive stigma surface of a pistil. Pollination also occurs in cone-bearing woody plants (pines, spruce, sequoias), but there is no pistil and the pollen grain is deposited directly onto a droplet made by an unfertilised seed (ovule) exposed to the air.

**Pollinator.** A dependable animal or natural force (air or water currents) that regularly and naturally transfers the viable pollen grains in anthers to the receptive stigmas in carpels/pistils of flowering plants. In cone-bearing plants (gymnosperms) pollinators deposit grains directly onto the pollination drops secreted by naked ovules.

**Pollinium (plural pollinia).** A mass of hundreds or thousands of pollen grains made in one chamber of an anther that stick together and are released as a cohesive lump or pellet. The pellet is usually attached to the viscidium by a stalk known as caudicle, stipe (in orchids) or translator arm (in members of the oleander/milkweed family (Apocynaceae)). As most anthers contain more than one pollen-making chamber, a single anther can produce two, four or more pollinia.

**Pooter or aspirator.** A small device invented by American entomologist, Frederick William Poos Jr. (1891–1987) for collecting the smallest insects. It allows the user to 'suck' the tiniest specimens into a collector's vial without inhaling or swallowing them.

**Proboscis.** Two or more mouthparts on an insect's face that interconnect to form a tube of varying lengths to consume fluids. A proboscis may come tipped with a sponge or spoon. Some proboscides are called stylets because they are so rigid and sharp they puncture plant or animal organs to drainthe host of sap or blood.

**Prodoxidae (prodoxid).** A family of moths known to pollinate *Yucca* flowers and many of the *Lithophragma* species of North America. The moths leave their eggs on or in a flower's ovary and the caterpillar eats the developing seeds.

**Proteaceae.** A family of woody plants largely confined to the Southern Hemisphere and most diverse in South Africa and Australia. It includes proteas, banksias and the gourmet macadamia nut.

**Protogyny.** Delayed sexual maturation in a bisexual (perfect) flower. The stigma is receptive to pollen when the flower opens but dries up or collapses hours or days before its anthers open and release their grains, preventing self-pollination. The opposite of protogyny is protandry where anthers release pollen first then collapse hours or days before their stigmas become receptive.

**Pteropodidae (pteropodid).** A large family of fruit- and nectar-feeding bats distributed from Africa east to tropical Asia, Australia and the islands of the South Pacific. They are often called flying foxes or fruit bats and lack much of the echolocation abilities associated with most bats in other families.

**Raphides.** Naturally forming, often needle-like crystals floating in the fluid of living plant cells.

**Reduction.** An evolutionary development in which structures or processes associated with an ancestor are reduced or lost over generations.

**Rostellum.** A specialised lobe on the stigma of most orchids. It does not receive incoming pollinia. Instead, the rostellum makes the sticky plug (viscidium) or glue secretions that becomes attached to the body of a pollinating animal as it leaves the flower. In some orchids, the rostellum also makes the connecting stalk known as the stipe.

**Rutaceae.** A family of more than 1,600 species of herbaceous and woody plants found worldwide except in the coldest regions of the Northern Hemisphere and Antarctica. Their vegetative and reproductive organs bear aromatic pellucid dots or glands. The family gives us citrus fruits and a number of commercial scents.

**Scabland.** A tough terrain of elevated, rocky regions with little topsoil that is intersected and subdivided by now dried up postglacial stream channels. Often too dry for woody plants, scablands may be rich in perennial wildflowers until the hot, dry summer sends them into dormancy.

**Sepal.** Usually the most leaf-like and outermost organs in a flower forming a part of the bud case and opening as a ring known as the calyx.

**Species.** One or more populations in which individuals share the same lifecycle and genetic ancestry.

**Sporopollenin.** A natural and versatile plant polymer. It is the primary substance that makes up the outer wall (exine) of most pollen grains. The sporopollenin molecule is believed to be based on repeating polymers of Vitamin A.

**Spur.** When applied to flowers it refers to a hollow, often conical or tubular 'bump out' on a petal or sepal that may or may not contain nectar for pollinators with elongated mouthparts.

**Stamen.** A male, pollen-making organ that makes up the whorl or spiral in a flower (the androecium). Each stamen usually consists of a pollen-making anther attached to a stalk-like sterile filament.

**Staminode.** A stamen that never produced pollen because it never developed an anther. Most orchids that make only one fertile stamen have two staminodes attached to either side of their columns. Sometimes the additional lobes made by some staminodes are called column wings or auricles.

**Staminodium.** The large shield-like and often sculptured, sterile stamen that sits in between two fertile anthers and hangs over the enlarged stigma in the column of a slipper orchid flower. In contrast, the majority of remaining orchid species make two staminodes, one on either side of a single, fertile anther.

**Steady-state flowering.** A plant that undergoes continuous flowering over weeks or months. Such plants make dozens or thousands of short-lived flowers, and only open a few flowers each day. Steady-state flowering encourages trapline pollinators.

**Stigma.** The tip or apex of a carpel or pistil. It becomes biochemically active when it is ready to receive and process pollen grains that adhere to its surface and/or secretions.

**Stipe.** A sterile stalk made by the sticky plug (viscidium) of an orchid flower that interconnects all the pollinia in the anther to the sticky plug, as in midge orchids (*Corunastylis*). The stipe and its viscidium aere made by the stigma lobe known as the rostellum.

**Style.** The 'neck' of a carpel or pistil that interconnects the stigma to the ovary. When pollen grains germinate, their pollen tubes containing sperm must first penetrate the stigma. These tubes grow down the style to reach the ovules in the ovary.

**Tepals.** Sterile organs in a flower that appear intermediate between sepals and petals but form rings or spirals like true sepals and petals.

**Thorax.** The mid-section of most insects containing the wings and legs. Many orchid flowers deposit their pollinaria on the back of an insect's thorax, in between the wings.

**Thynnidae.** A family of solitary wasps with a centre of diversity in the Southern Hemisphere, especially Australia and South America. In many species, males are winged and larger than the wingless females that lay their eggs in beetle grubs that feed on roots.

**Trapline foragers/pollinators.** Flying animals that cover broad areas in search of food by remembering those plants that offer limited but consistent rewards (usually sugar-rich nectar) during each period of foraging activity. Trapline foragers are often dependent on steady-state flowering plant species.

**Trophallaxis.** The mutual exchange of food by regurgitation between adults and between adults and their offspring or siblings. It is especially common in social insects like honeybees, ants, termites and certain wasps.

**Umwelt.** The range of senses an animal employs to perceive environmental cues. The strength of an umwelt differs between species. For example, a dog's perception of colour and scent to those of a human.

**Ur-pollinator.** The ancestral and extinct groups of animals that pollinated the first true flowers. Prospective ur-pollinators, based on fossil evidence, include an increasing number of beetles, flies, moths with descendants in modern families and specimens appearing intermediate between bees and wasps.

**Viscidium.** The sticky plug made by a lobe (rostellum) on the stigma of an orchid. The vascidium may be made of a natural glue or it is a detachable plug of shed cells. In most orchids the viscidium becomes attached to the pollinator as it exits the flower.

# Annotated bibliography and additional reading

We are fortunate to live in an era in which most scientific papers become available online within a few years after they are published and I have provided links when possible. Books on the plant sciences are a different matter and may be more difficult to access in the collections of most public libraries. Readers are encouraged to request interlibrary loans or consult speciality libraries in museums, botanical gardens, arboreta and universities with departments of biology, agriculture and/or conservation.

## INTRODUCTION

Ballantine B (1968) *Nobody Loves a Cockroach... Or Fly, Ant, Bat, Rat, or Other Creeping, Crawling, Flying Pests Which Menace Our Daily Lives*. Little, Brown and Company, USA. It is worth finding if only to see how much and how little our attitudes have changed about pests and their eradication. Does industrial chemistry always provide the best answer?

Bernhardt P, White D (2022) In memoriam. Leonard Thien. *Plant Science Bulletin* **68/1** (Spring), 81–84. <https://botany.org/userdata/IssueArchive/issues/originalfile/WebPSB68_1_2022_3.pdf>. The article also contains colour photos of some of the flowers he studied, including a star anise (*Illicium*) in bloom.

Wilson EO (1987) The little things that run the world (the importance and conservation of invertebrates). *Conservation Biology* 1/**4**, 344–346. <https://www.jstor.org/stable/2386020?seq=3>. Everyone contemplating a career in the life sciences needs to read this, even if you see yourself as lab oriented and all you want to do is add or delete genes to a few specialised cells. Wilson's contention is that there is no planet Earth, as we know it, without invertebrates consuming cellulose, dung and filtering muddy sediments on land and sea.

## CHAPTER 1. WHY AND HOW DO FLOWERS VARY?

Barth FG (1991) *Insects and Flowers: The Biology of a Partnership*. Princeton University Press, New Jersey. Firmly recommended if you want a simple and direct introduction explaining how insects perceive flowers and exploit floral resources. In particular, there is a labelled diagram showing how insect mouthparts are modified to take up fluids and small particles. It compares the parts of a cockroach's 'jaws and tongue' to those of a honeybee, butterfly and housefly.

Bernhardt P (2002) *The Rose's Kiss: A Natural History of Flowers.* University of Chicago Press, Illinois. Think of this as a 'popular mechanics' of how flowers work and how they interact with different parts of their environment. After teaching pollination biology to undergraduates for 15 years I thought I had enough information to explain the subject to more people. Note: the final chapter on fossil flowers is now out of date due to the recent work on the fossils in Myanmar amber.

Dörken VM, Edwards D, Ladd PG, Parsons RF (2021) *The Four Dimensions of Terrestrial Plants: Reproduction, Structure, Evolution and Ecology*. Kessel Publishing House, Remagen, Germany. This book was written by botanists for botanists and uses a much more specific and technical language compared to a college textbook (see Evert & Eichhorn, below). It is worth the effort as two authors are Australian, so plant diversity in the southern hemisphere is not ignored, and there are excellent photos comparing reproductive structures in gymnosperms versus flowering plants.

Evert RF, Eichhorn SE (2013) *Raven Biology of Plants.* 8th edn. W.H. Freeman and Company, New York. Everyone interested in plant structure and their lifecycles needs a book written to the level of university undergraduates. As it contains hundreds of colour photos and micrographs, you learn a lot if you only read the photo captions.

Fahn A (1979) *Secretory Tissues in Plants*. Academic Press, New York. The chapter on nectar glands is recommended as it shows how they vary in location and shape in the flowers of different species. You will also see how nectar glands interconnect with the internal threads of sugar-conducting tissues, known as phloem.

Gomez B, Daviero-Gomez V, Coiffard C, Dilcher DL (2015) *Montsechia*, an ancient aquatic angiosperm. *Proceedings of the National Academy of Sciences* **112/35**, 10985–10988. doi:10.1073/pnas.150924111. Read how the fossil of one of the tiniest flowers was found and documented (the photos are surprisingly clear). The description of this extinct water weed changes our understanding of the evolution and relationships among lineages of flowering plants.

Kaiser R (2010) *Scent of the Vanishing Flora.* Wiley-VCH, Zurich. Don't miss this one. At 400 pages, it is a deep and intense crash course into the biochemistry of flower, leaf and wood fragrances using some of the rarest species in the world.

Lake M (1989) *Scents and Sensuality: The Essence of Excitement*. John Murray Publishers, London. An Australian classic, and a personal favourite, offering a brief history of what we smell and taste and why we like it.

Lee D (2007) *Nature's Palette: The Science of Plant Colour*. University of Chicago Press, Illinois. Possibly the best introduction to the effects plants employ to produce so many visual patterns despite the limited number of plant pigments. Includes an excellent illustration of cone cells in the eyes of very different animals.

Poinar G (2020) Discoscapidae fam nov. (Hymenoptera: Apidae), a new family of stem lineage bees with associated beetle triungunlins in mid-Cretaceous Burmese amber. *Palaeodiversity* 13/1, 1–9. doi:10.18476/pale.v13.a1. It's so easy to become a fan of Dr Poinar's books and papers describing organisms in amber because he eventually puts all the data together to give us a vision of a long extinct habitat. The colour photographs show the pollen the bee carried before it perished in tree resin.

Poinar G, Poinar R (1999) *The Amber Forest: A Reconstruction of a Vanished World.* Princeton University Press, New Jersey. The book is for beginners. It helps you understand what science learns from the remains of extinct tiny things. With a large amber deposit, you can piece together an extinct ecosystem that flourished 15–45 million years ago.

Vogel S (1990) *The Role of Scent Glands in Pollination*. Routledge, London. While Kaiser (above) introduces you to the diversity of molecules that make up plant scents, Vogel shows you where they are found on the flower, their various shapes and when the cells are ready to release their contents. This is the same man who found microscopic, nectar-secreting pits on some flowers and named them nectarioles.

## CHAPTER 2. GETTING UGLY WITH THE GODDESS

Bernhardt P (2016) Surviving a decade of drought? Some *Cypripedium* species of Lijiang, Yunnan. *Orchids* **85/2** (February), 122–125. <https://www.researchgate.net/profile/Peter-Bernhardt-2/post/Does_anybody_know_any_study_with_succulent_flowers/attachment/59d62d8779197b807798bcce/AS%3A350685704015873%401460621395605/download/8502-pb-v2.pdf>. Here is an opportunity to look at more of the slipper orchid species of China. Climate change has reached the Heng Duan Mountains.

Edens-Meier R, Bernhardt P (Eds) (2014) *Darwin's Orchids: Then & Now*. University of Chicago Press, Illinois. Twenty-seven authors worked together to produce 12 interrelated chapters. The first chapter traces the history of Darwin's decision to study orchid flowers and its impact on evolutionary biology. Chapter 10 is a close look at pollination in the subfamily Cypripedioideae, which includes lady's slippers in the genus *Cypripedium*.

Edens-Meier R, Arduser M, Camilo GR, Tackett MJ (2021) Pollination ecology and breeding systems of *Cypripedium kentuckiense* (Orchidaceae) in Tennessee. *Journal of the Torrey Botanical Society* **148/1**, 53–74. doi:10.3159/TORREY-D-20-00033.1. This research shows how much work still needs to be done and what we need to learn as we take steps to protect endangered species. Warning: If you follow Dr Edens-Meier's YouTube channel it also shows that people studying native orchids in America must be wary of bears.

Jiang H, Kong J-J, Chen H-C, Xiang Z-Y, Zhang W-P, Han Z-D, Liao P-C, Lee Y-I (2020) *Cypripedium subtropicum* (Orchidaceae) employs aphid colony mimicry to attract hoverfly (Syrphidae) pollinators. *New*

*Phytologist* **227/4**, pp. 1213–1221. doi:10.1111/nph.16623. A paper of this quality shows you the intense detective work, applied microscopy and forensic techniques needed to interpret what is going on between flowers and insects.

## CHAPTER 3. ONE HUNDRED YEARS OF MOSQUITOES AND ORCHIDS

Lahondère C, Vinauger C, Okubo RP, Riffel JA (2019) The olfactory basis of orchid pollination by mosquitoes. *Proceedings of the National Academy of Science* **117/1**, 708–716. doi:10.1073/pnas.191058911. This study compares the scents of the blunt-leaved orchid to other North American members of the genus *Platanthera.* Imagine all the investigations conducted on something as miniscule as individual segments of a mosquito's antennae. Consider its future implications for public health.

Thien LB, Utech F (1970) The mode of pollination in *Habenaria obtusata* (Orchidaceae). *American Journal of Botany* **57/9**, 1031–1035. doi:10.1002/j.1537-2197.1970.tb09905.x. This paper looks so simple compared to the previous two, but is it? Think of how much Len accomplished by himself with simple tools. Remember, taxonomists reorganised the genus *Habenaria,* and placed North American species into the genus *Platanthera* years after this paper was published.

## CHAPTER 4. BLOOD OR TEARS?

Bernhardt P, Edens-Meier R, Grimm W, Ren Z-X, Towle B (2017) Global collaborative research on the pollination biology of rare and threatened orchid species (Orchidaceae). *Annals of the Missouri Botanical Garden* **102/2**, 364–376. doi:10.3417/D-16-00005A. Do specialised pollination systems in the orchid family represent the most vulnerable link in their lifecycles? It's an important question as people degrade their environments and climate change alters flowering and fruiting seasons.

Calder M (1998) John Roslyn (Ros.) Garnett, AM 1906–1998. *The Victorian Naturalist* **115/2**, 70–71. <https://www.biodiversitylibrary.org/item/139549#page/80/mode/1up>. Here is a tribute to the life of another amateur naturalist of the 20th century who made a significant contribution to our understanding of how the Australian flora and fauna work together.

Garnet JR (1940) Observations on the pollination of orchids. *The Victorian Naturalist* **56**, 191–197. <https://www.biodiversitylibrary.org/item/126466#page/209/mode/1up>. A great example of what an earlier Australian generation did to illuminate the lifecycles of unfamiliar and understudied species. The author turned his hobby into exploratory research with simple tools, photos and pen and ink drawings.

Ren Z-X, Grimm W, Towle B, Qiao Q, Bickel DJ, Outim SK, Bernhardt P (2023) Comparative pollination ecology, fruit set and seed set in *Corunastylis* species (Orchidaceae). *Plant Systematics and Evolution* **309**, 7. doi:10.1007/s00606-023-01845-3. The irony is obvious. Garnet worked alone over several years while we needed a gang to cover the fieldwork, microscopy and insect identification.

## CHAPTER 5. GNATS IN A WINTER BOUQUET

Bernhardt P (1995) Notes on the anthecology of *Pterostylis curta* (Orchidaceae). *Cunninghamia* **4**, 1–8. <https://www.botanicgardens.org.au/sites/default/files/2023-09/Volume-4%281%29-1995-Bernhardt1-8.pdf>. I include this very slight example of my work as an example of why fieldwork isn't always enough and you need controlled conditions to answer some questions. Field and greenhouse botanists need to work together.

Bernhardt P (2008) *Gods and Goddesses in the Garden: Greco-Roman Mythology and the Scientific Names of Plants.* Rutgers University Press, New Jersey. The men who gave us the system we use today to give scientific names to plants had classical educations and knew the Greco-Roman myths. Assigning plant genera, species, some families and their organs to the names of half-forgotten deities, heroes and monsters lets us retell the old stories.

Blanco MA, Barboza G (2005) Pseudocopulatory pollination in *Lepanthes* (Orchidaceae: Pleurothallidinae) by fungus gnats. *Annals of Botany* **95/5**, 763–772. doi:10.1093/aob/mci090. This is one of the seminal papers about where a gnat will leave its semen. The big difference here is that the lip petal of a babyboot orchid (*Lepanthes*) will bring the insect to climax without imprisoning him.

Jones DL (2021) *A Complete Guide to Native Orchids: A Botanical Obsession: The Legacy of a Lifetime's Study of Australian Orchids*. 3rd edn. Reed New Holland, Wahroonga, NSW. Possibly the most spectacular and expensive book anyone has ever written on the natural history of a family of flowering plants found in Australia. At least 1,300 orchid species are described and illustrated with colour photos and/or technical drawings.

Kuiter R, Findlater-Smith MJ (2017) 'Initial observations on the pollination of *Corybas* (Orchidaceae) by Fungus-gnats (Diptera: Sciaroidea)'. Short Paper 5. Aquatic Photographics, Seaford, Vic. Rudie Kuiter prefers to publish his own research online. The photos in this paper show what can be done with extreme patience, the best equipment and a lot of patience when the weather is uncooperative.

Singh AU, Sharma K (2016) Pests of mushroom. *Advances in Crop Science Technology* **4/2**, 213. doi:10.4172/2329-8863.1000213. The range and destructive capacity of fungus gnats on an expanding number of edible fungi may surprise you. It also covers nematodes (eelworms), mites, springtails and flies.

## CHAPTER 6. WASPISH

Bernhardt P, Meier R, Vance N (2013) Pollination ecology and floral function of Brown's peony (*Paeonia brownii*) in the Blue Mountains of northeastern Oregon. *Journal of Pollination Ecology* **11/2**, 9–20. doi:10.26786/1920-7603(2013)2. This is one of two papers my consortium published based on fieldwork completed at GROWISER, a wildflower refuge. We needed additional help from entomologists to identify our wasps and biochemists to analyse the sugar and amino acid levels in the flower's nectar.

Cohen D, Siegel A (2021) *Ashkenazi Herbalism: Rediscovering the Herbal Traditions of Eastern European Jews.* North Atlantic Books. Berkley, California. This is where I found the atypical use of peony petals as a medicine for female complaints.

Schlising RA (1976) Reproductive proficiency in *Paeonia californica* (Paeoniaceae). *American Journal of Botany* **63/8**, 1095–1103. doi:10.1002/j.1537-2197.1976.tb13194.x. Currently, there is only one publication for those interested in the reproduction of the second peony species in North America.

## CHAPTER 7. A CORSAGE FOR WASPS

Bernhardt P (2016) A natural history of pollination at a shrine: orchids in Yunnan. *Orchids* **82/10**, 764–767. Wasps pollinate *Epipactis* orchids all over Europe, but the site and circumstances under which I observed the process in China were unique. The article also gives us a brief look into the culture and religion of the Nakhi people.

Coleman E (1927) Pollination of the orchid Cryptostylis leptochila. The Victorian Naturalist 44/1, 20–22. <https://www.biodiversitylibrary.org/item/324327#page/38/mode/1up>. Everyone interested in orchid flowers and wasps in Australia needs to start here while understanding that Coleman devoted her observations to more tongue, bonnet and duckbill orchid species from 1928 to 1938, which appeared in later publications. She also studied some native Australian orchids pollinated by bees.

Dixon KW, Tremblay RL (2009) Biology and natural history of *Caladenia*. *Australian Journal of Botany* **57/4**, 247–258. doi:10.1071/BT08183. As *Caladenia* is the largest genus of orchids in Australia they are an excellent model for understanding the process of adaptive radiation. This paper allows you to compare food and sex mimics, but also shows that nectar production and sex mimicry overlap in some spider orchids.

Kuiter R (2015) *Orchid Pollinators of Victoria*. 3rd edn. Aquatic Photographics, Seaford, Vic. This book allows you to view the most intimate photographs of male wasps visiting flowers and carrying the pollinia of members of the genera *Caladenia*, *Cryptostylis*, *Chiloglottis* and *Prasophyllum* in closest detail. It's the only book offering colour photos showing how winged males of the pygmy bulldog ant (*Myrmecia urens*) pollinate the hare orchid (*Leporella fimbriata*).

## CHAPTER 8. BEAUTY THROUGH A BEETLE'S EYES

Goldblatt O (1987) *The Moraeas of Southern Africa: A Systematic Monograph of the Genus in South Africa, Lesotho, Swaziland, Transkei, Botswana, Namibia and Zimbabwe*. National Botanic Gardens with the Missouri

Botanical Garden, USA. The remarkable watercolours of Fay Anderson allow the reader to compare the flowers of beetle-pollinated peacock moraeas to other members of the genus.

Goldblatt P, Bernhardt P, Manning JC (1998) Pollination of petaloid geophytes by monkey beetles (Scarabaeidae: Rutelinae: Hopliini) in southern Africa. *Annals of the Missouri Botanical Garden* **85/2**, 215–230. This paper presents the results of my first attempt to remove and identify the pollen carried by hundreds of monkey beetles collected and pinned in South Africa by Drs Goldblatt and Manning.

Goldblatt P, Manning JCG (2008) *The Iris Family: Natural History and Classification*. Timber Press, Portland, Oregon. Describes and illustrates all the genera in this family while allowing the reader to focus on its diversity in southern Africa. The natural history section compares beetle-pollination in this family to other pollination systems. It also addresses broader topics including the response of plants to fire, variation in their underground stems, how they disperse their seeds and aspects of their lifecycles.

Picker MD, Midgley JJ (1996) Pollination by monkey beetles (Coleoptera: Scarabaeidae: Hopliini): flower and colour preferences. *African Entomology* **4/1**: 7–14. <https://journals.co.za/doi/abs/10.10520/AJA10213589_177>. I consider this a seminal study. The way in which the authors tested and investigated their hypotheses served as templates for the field and garden-based experiments of future scientists working on similar topics.

## CHAPTER 9. NATIVE ROSE AND ITS INVISIBLE WORM

Bussell BM, Considine JA, Spader ZE (1995) Flower and volatile oil ontogeny in *Boronia megastigma*. *Annals of Botany* **76/5**, 457–463. doi:10.1006/anbo.1995.1120. The paper works at two levels. First, where are the glands located in the flower and, second, how do specific scent molecules vary in volume as the flower ages?

Costermans L (2009) *Native Trees and Shrubs of South-eastern Australia: Covering Areas of New South Wales, Victoria and South Australia*. New Holland, Sydney. The first edition (1981) remains a sentimental favourite that helped teach me to identify the wealth of Australia's woody plants. Using the keys, drawings, maps and photos to the new edition will make it easier for you to discriminate species of boronias from wax flowers (*Eriostemon*), croweas (*Crowea*) and stinkwood (*Zieria*).

Milla E (2019) *Evolution and Ecology of the Australian Heliozelidae* (Adeloidea, Lepidoptera). PhD thesis. Faculty of Science, The University of Melbourne, Victoria. <https://minerva-access.unimelb.edu.au/items/03bc0ff0-ed8b-5bdc-9514-8edb9591c399>. The thesis is so well illustrated you will have no problems seeing the fine sculptures on the flowers and the moths.

Pellmyr O (2003) Yuccas, yucca moths, and coevolution: a review. *Annals of the Missouri Botanical Garden* **90/1**, 35–55. A most thorough treatment of the topic taking us from the second half of the 19th century into the early 21st and written by an authority in the field. I attended the conference on which this publication was based and remember how the audience gasped at the footage of the moth's tentacles forming pollen into balls.

## CHAPTER 10 WHEN BRUTAL BATS BITE BIG BUDS

Bernhardt P (1995) The bat in the hat: flying foxes in Australian children's literature. *Bats* **13/4**, 15–17. During the 20th century, a few Australian authors and illustrators presented their young readers with flying fox characters that were harmless, funny and heroic.

Corner EJH (1988) *Wayside Trees of Malaya*. Volumes I and II. 3rd edn. Malayan Nature Society, Kuala Lumpur. Written by one of the greatest students of tropical botany of the 20th century. Corner's descriptions of tree architecture, seasonality and lifecycles are as surprising as they are informative.

Cox PA, Sacks OW (2002) Cycad neurotoxins, consumption of flying foxes, and ALS-PDC disease in Guam. *Neurology* **58/6**, 956–959. doi:10.1212/WNL.58.6.95. This is the paper that attempted to change the way medical researchers approached this devastating disease. While this condition may vanish from Guam, Dr Cox continues to work on isolating the molecules that may serve as model systems for the modern rogues' gallery of neurological disorders.

Fleming TH, Kress WJ (2013) *The Ornaments of Life: Coevolution and Conservation in the Tropics*. University of Chicago Press, Illinois. An exhaustive and meticulous study of evolutionary ecology focusing primarily on the birds and mammals that pollinate tropical trees and/or disperse their seeds. Includes easy to understand graphs and some fine colour photos.

Graff S (2017) Guam's "skeleton key" enigma. *Penn Medicine News*. <https://www.pennmedicine.org/news/publications-and-special-projects/penn-medicine-magazine/archived-issues/2017/winter-2017/guam>. Addresses the full history of the neurodegenerative diseases among the Chamorro, including a heartbreaking case study and the history of repeated attempts to find the causes. Note that it dismisses the work of Cox and Sacks.

Hutson T (2022) *Bats: Their Biology and Behaviour*. Natural History Museum, London. Chapter 3 includes a short factual essay comparing the diversity, anatomy and foraging habits of New World leaf-nose bats and Old World flying foxes.

## CONCLUSIONS

Bezerra ELS, Machado IC, Mello MAR (2009) Pollination networks of oil-flowers: a tiny world within the smallest of worlds. *Journal of Animal Ecology* **78/5**, 1096–1101. doi:10.1111/j.1365-2656.2009.01567.x. This is the simplest example (see Figure 1A) of a pollination network I could find as some researchers now use AI to make their networks. It shows the interdependence of related species of oil-collecting bees to oil-secreting flowers of members of the nance or Barbados cherry family (Malphighiaceae) found in a semi-arid thorn forest in Brazil.

Ollerton J (2021) *Pollinators & Pollination: Nature and Society*, Pelagic Publishing, Exeter, UK. If you want to study pollination you can learn the basics here. It was written with a European audience in mind, but the techniques are the same and there are useful tips about tools including one found in your laundry.

Sidman J (2018) *The Girl Who Drew Butterflies: How Maria Merian's Art Changed Science,* Houghton Mifflin Harcourt, Boston. This is a children's book, but how many children's books have a glossary and show you how ground up lapis lazuli made the best blue paint? Adults should read it before offering it to kids as the history of this artist focuses on how illustration best served science long before photography.

# General index

# Taxonomic index